Encyclopaedia of Mathematical Sciences

Volume 42

Editor-in-Chief: R. V. Gamkrelidze

V. P. Khavin N. K. Nikol'skiĭ (Eds.)

Commutative Harmonic Analysis IV

Harmonic Analysis in $\mathbb{R}^n$

Springer-Verlag
Berlin Heidelberg New York
London Paris Tokyo
Hong Kong Barcelona
Budapest

Title of the Russian edition:
Itogi nauki i tekhniki, Sovremennye problemy matematiki,
Fundamental'nye napravleniya, Vol. 42, Kommutativnyĭ garmonicheskiĭ analiz 4
Publisher VINITI, Moscow 1989

Mathematics Subject Classification (1991):
31-XX, 42-XX, 43-XX, 46-XX

ISBN 978-3-642-08103-3

Printed in the United States of America
41/3140-543210 - Printed on acid-free paper

Preface

In this volume of the series "Commutative Harmonic Analysis", three points mentioned in the preface to the first volume are realized: 1) Multiple Fourier series and Fourier integrals; 2) The machinery of singular integrals; 3) Exceptional sets in harmonic analysis.

The first theme is the subject matter of the contribution by Sh. A. Alimov, R. R. Ashurov, A. K. Pulatov, which in an obvious way constitutes the "multidimensional parallel" to S. V. Kislyakov's article in Volume I, devoted to the "inner" questions of Fourier analysis of functions of one variable. The passage to the analysis of functions defined on $\mathbb{R}^n$, $n > 1$, tells us something essential about the nature of the problem under study.

The contribution by E. M. Dyn'kin, the beginning of which was already published in Volume I of this subseries, is devoted to singular integrals. Besides classical material (Calderón-Zygmund and Littlewood-Paley theory), this article contains an exposition of recent results, which in an essential way have widened the scope of the whole area and have made it possible to solve many old problems, thereby sometimes transcending the very frames of harmonic analysis in its canonical interpretation.

Quite different but highly interesting and often tantalizing material is collected in S. V. Kislyakov's contribution, which concludes this volume, centering around the notion of "exceptional" (or "narrow") sets (this topic was briefly discussed already in V. P. Khavin's contribution in the first volume of this series). Special attention is given to the so-called Sidon set, which are connected with some important results obtained in recent years, and further to methods utilizing the notion of capacity, which has its origin in potential theory.

V. P. Khavin, N. K. Nikol'skiĭ

List of Editors, Authors and Translators

Editor-in-Chief

R. V. Gamkrelidze, Academy of Sciences of the USSR, Steklov Mathematical Institute, ul. Vavilova 42, 117966 Moscow, Institute for Scientific Information (VINITI), ul. Usievicha 20a, 125219 Moscow, USSR

Consulting Editors

V. P. Khavin, Department of Mathematics, Leningrad University, Staryi Peterhof, 198094 Leningrad, USSR

N. K. Nikol'skiĭ, Steklov Mathematical Institute, Fontanka 27, 191011 Leningrad, USSR

Authors

Sh. A. Alimov, Tashkent State University, Vuzgorodok, 700095 Tashkent, USSR

R. R. Ashurov, Tashkent State University, Vuzgorodok, 700095 Tashkent, USSR

E. M. Dyn'kin, Department of Mathematics, Leningrad Institute of Electrical Engineering, Prof. Popov st. 5, Leningrad, USSR

S. V. Kislyakov, Steklov Mathematical Institute, Fontanka 27, 191011 Leningrad, USSR

A. K. Pulatov, Tashkent State University, Vuzgorodok, 700095 Tashkent, USSR

Translator

J. Peetre, Matematiska institutionen, Stockholms universitet, Box 6701, S-113 85 Stockholm, Sweden

Contents

I. Multiple Fourier Series and Fourier Integrals

Sh. A. Alimov, R. R. Ashurov, A. K. Pulatov

Translated from the Russian
by J. Peetre

Contents

Introduction

1. How Multiple Trigonometric Series Arise. The basic objects of study in this article are functions of several variables which are periodic in each of these variables. We may assume (this assumption is not very restrictive) that the corresponding periods are the same and equal to 2π. Thus, we may take the fundamental set, where our functions are defined, to be the N-dimensional cube

$$\mathbb{T}^N = \{x \in \mathbb{R}^N : -\pi < x_j \le \pi, j = 1, \dots, N\}.$$

This cube can be identified in a natural way with an N-dimensional torus (i. e. the subset of $\mathbb{C}^N$ of the form $(e^{ix_1}, \dots, e^{ix_N})$, where $(x_1, \dots, x_N) \in \mathbb{R}^N$), which in what follows we shall not distinguish from $\mathbb{T}^N$.

The fundamental harmonics with respect to which we take our expansion have in appearance exactly the same form as in the one-dimensional case: e^{inx} where, for $N > 1$, nx stands for the inner product

$$nx = n_1x_1 + \dots + n_Nx_N\,, \quad n \in \mathbb{Z}^N, x \in \mathbb{R}^N.$$

Here, as usual, $\mathbb{Z}^N$ is the set of all vectors with integer components: $n = (n_1, \dots, n_N)$ is in $\mathbb{Z}^N$ if $n_j = 0, \pm 1, \dots$.

A *multiple trigonometric series*

$$\sum_{n\in\mathbb{Z}^N} c_n e^{inx}, \tag{1}$$

where the coefficients c_n are arbitrary complex numbers, looks exactly as an ordinary "one-dimensional" series. This exterior resemblance leads one in a completely natural way to the idea that the multidimensional version of Fourier analysis is subject to the general principles which, in particular, were discussed in the introductory article of the series "Commutative Harmonic Analysis" (Khavin (1988)). We run into a similar situation when we consider the vibrations of a string, which is subject to the same principles as the vibrations of complicated spatial objects. This analogy, which may appear superficial, has in fact a deep interior meaning.

Multidimensional harmonic analysis largely owes its appearance to the study of various vibrating systems. As is well-known, the early 19th century witnessed new progress in the mathematical description of physical processes taking place in real space, without assumptions of symmetry, which often reduce the situation to functions of one variable. In the first place let us here mention the classical work of Fourier in the theory of the distribution of heat (1807, the detailed study appearing in 1822), as well as the work of Laplace and Dirichlet, where not only the celebrated "method of Fourier" was set forth but, essentially, also a considerable portion of the problems of harmonic analysis were formulated which came to determine the development of this subject for a long period of years.

Trigonometrical series on the real line $\mathbb{R}^1$ may be found already in the 18th century in the work of Euler, D. Bernoulli and Clairaut – mainly in connection with the vibrating string. Now the era of "multidimensional" problems starts – it concerns the distribution of heat and the vibrations of complicated three-dimensional bodies. In the above mentioned work of Fourier the first attempts were made to solve the problem of heat distribution in spatial bodies, in particular, for the cube $\mathbb{T}^3$ by reducing it to the problem of expanding an arbitrary function in three variables $f(x_1, x_2, x_3)$ in a triple sine series, i. e. to the problem of representing a given function in the form of an infinite series

$$f(x) = \sum_{n_1=1}^{\infty} \sum_{n_2=1}^{\infty} \sum_{n_3=1}^{\infty} b_{n_1 n_2 n_3} \sin n_1 x_1 \sin n_2 x_2 \sin n_3 x_3 .$$

The *method of Fourier*, applied with success in mathematical physics under the course of two centuries, includes a *separation of variables*. Perhaps, nothing describes it better than the renowned saying "divide and rule" (*divide et impera*). One of the first problems which were effectively solved by Fourier's method was the problem of transversal vibrations of a plane membrane Ω, depending only on the expansion, and not on the curvature. The vertical distortions $u = u(x_1, x_2, t)$ of a point of the membrane with coordinates (x_1, x_2) satisfy the equation

$$\frac{\partial^2 u}{\partial t^2} = c^2 \Delta u, \quad x \in \Omega, t > 0, \tag{2}$$

where $\Delta = \frac{\partial^2}{\partial^2 x_1^2} + \frac{\partial^2}{\partial^2 x_2^2}$ is the *Laplace operator*. If the boundary $\partial\Omega$ of the membrane is stiffly fastened, then we have $u = 0$ along the boundary curve for $t \geq 0$. Usually one assumes that the position and the speed of each point of the membrane are given for $t = 0$:

$$u\big|_{t=0} = f, \quad \frac{\partial u}{\partial t}\Big|_{t=0} = g. \tag{3}$$

The method of separation of variables consists of first, neglecting the conditions (3), seeking a solution of equation (2) which is periodic in t and just satisfies the boundary conditions (simple harmonics):

$$u_\nu = v_\nu (a_\nu \cos \nu t + b_\nu \sin \nu t),$$

which leads to the following boundary problem for the eigenvalues and eigenfunctions of the Laplace operator:

$$\Delta v + \lambda v = 0, \quad x \in \Omega, \ \lambda = \frac{\nu^2}{c^2},$$
$$v\big|_{\partial\Omega} = 0.$$

When the eigenfunctions $v_\nu(x)$ (and thus the "spectrum of the problem", i. e. the set of λ's for which the above harmonic oscillations with frequency $\nu = c \cdot \sqrt{\lambda}$ do exist) are found, we obtain by virtue of the superposition principle, a consequence of the linearity of our problem, the solution of equation (2) subject to zero boundary conditions as the sum of the series

$$u(x,t) = \sum_\nu v_\nu(x)(a_\nu \cos \nu t + b_\nu \sin \nu t).$$

Now we can turn to the initial conditions (3) and try to satisfy them by adjusting the coefficients a_ν and b_ν, that is, we have to choose them in a such a way that the following identities are fulfilled:

$$f(x) = \sum_\nu a_\nu v_\nu(x), \quad g(x) = \sum_\nu \nu b_\nu v_\nu(x).$$

It is clear that the eigenfunctions $v_\nu(x)$ depend on the form of the membrane Ω, while it follows from the general spectral theory of operators that the functions v_ν form a complete orthonormal system in $L_2(\Omega)$, which allows us to find the coefficients in the expansion by the formulae

$$a_\nu = \int_\Omega f(x) v_\nu(x) dx, \quad b_\nu = \frac{1}{\nu} \int_\Omega g(x) v_\nu(x) dx.$$

In the case of a rectangular membrane

$$\Omega = \{x \in \mathbb{R}^2 : 0 < x_1 < \pi, 0 < x_2 < \pi\}$$

the variables separate also in the eigenvalue problem and we obtain (following Fourier)

$$v_\nu(x) = \frac{2}{\pi} \sin n_1 x_1 \sin n_2 x_2,$$

$$\nu = c\sqrt{n_1^2 + n_2^2}, \quad n_j = 1, 2, \dots, j = 1, 2.$$

The expansion of f then takes the form

$$f(x) = \sum_{n_1=1}^{\infty} \sum_{n_2=1}^{\infty} f_{n_1 n_2} \sin n_1 x_1 \sin n_2 x_2,$$

where

$$f_{n_1 n_2} = \frac{4}{\pi^2} \int_0^{2\pi} \int_0^{2\pi} f(x_1, x_2) \sin n_1 x_1 \sin n_2 x_2 \, dx_1 dx_2.$$

The function $g(x)$ admits an analogous expansion. It is not hard to see that these expansions are special cases of double trigonometric series.

2. What Do We Mean by the Sum of the Series (1)? The pecularities of the analysis on the tori $\mathbb{T}^N$ in higher dimensions ($N > 1$), distinguishing it from the one-dimensional situation, become visible at the very first attempt to find the sum of the series (1). Of course, by the sum of (1) one has to understand the limit of *partial sums*, sums of a finite number of terms of the series.

However the specific character of the multidimensional case shows itself in the great variety of "natural" definitions of partial sums.

The situation is similar to the one which arises when we try to define a multiple generalized integral over the entire space $\mathbb{R}^N$. If we are permitted to select the expanding sequence of connected domains exhausting $\mathbb{R}^N$ arbitrarily, one can show that the convergence of the multiple integral is equivalent to its absolute convergence.

Exactly in the same way, if we take arbitrary expanding sequences of bounded sets Ω_h exhausting $\mathbb{R}^N$ and define the partial sums of the series (1) by the formula

$$S_h = \sum_{n \in \Omega_h \cap \mathbb{Z}^N} c_n e^{inx},$$

then only those series converge which are absolutely convergent. However, the class of absolutely convergent series, despite their importance, is by far too narrow. As each absolutely convergent trigonometric series is uniformly convergent, this class is automatically contained in the space $C(\mathbb{T})$ of all continuous 2π-periodic functions but does not exhaust the latter.

Therefore, in order to be able to extend further the class of multiple series, to include series whose sums should have a completely determined meaning, one has to restrict in a reasonable way the sets Ω_h over which one extends the indices of the terms of the series entering in the partials sums. Resorting once more to the analogy with multiple integrals we may say that the nonexistence of unconditionally convergent integrals nevertheless produces integrals which are convergent in the sense of "principal values" (i. e. in the sense of limits of integrals taken over expanding regular sets such as balls, cubes, ellipsoids or parallelotopes, with restrictions on their halfaxes or side lengths).

In order to be able to better discuss the various methods of defining partial sums, to be set forth below, let us first consider the one-dimensional case.

As is well-known, the classical (one dimensional) trigonometric series

$$\frac{a_0}{2} + \sum_{n=1}^{\infty}(a_n \cos nx + b_n \sin nx)$$

can be written in the complex form (1) by putting

$$c_n = \tfrac{1}{2}(a_n - ib_n),\ c_{-n} = \bar{c}_n.$$

The "natural" partial sum

$$S_k = \frac{a_0}{2} + \sum_{n=1}^{k}(a_n \cos nx + b_n \sin nx)$$

then takes the form

$$S_k = \sum_{|n|\le k} c_n e^{inx}, \tag{4}$$

i. e. it leads indeed to taking the sum in principal value sense. This definition generalizes to the multidimensional case in three obvious manners, each leading to the following three forms of partial sums, which mainly will be encountered within this article.

3. Various Partial Sums of a Multiple Series. A) The *rectangular partial sum* $S_m(x)$ of the series (1) is defined by a vector $m \in \mathbf{Z}^N$ with nonnegative coordinates and takes the form

$$S_m(x) = \sum_{|n_1|\le m_1} \sum_{|n_2|\le m_2} \cdots \sum_{|n_N|\le m_N} c_n e^{inx}. \tag{5}$$

We say that the series (1) is *rectangularly convergent* (or *convergent in the sense of Pringsheim*) if the limit of the partial sums (5) exists for $\min m_j \to \infty$.

Despite the apparent "naturality" of this definition, it is not deprived of paradoxes. One such paradox is that there exists an uncountable set of partial sums which can not be forced to convergence in any way. As an example we may take the double series

$$\sum_{n_1=-\infty}^{\infty} \sum_{n_2=-\infty}^{\infty} c_{n_1 n_2} e^{i(n_1 x_1 + n_2 x_2)} \tag{6}$$

with coefficients $c_{n_1 n_2} = n_1^2(\delta_{n_2 0} - \delta_{n_2 2})$, where δ_{jk} is the Kronecker symbol. The rectangular partial sums of this series at the point (0,0) take the form

$$S_{m_1 m_2}(0) = \sum_{k=-m_1}^{m_1} \sum_{l=-m_2}^{m_2} k^2(\delta_{l0} - \delta_{l2}).$$

It is clear that $S_m(0) = 0$ for $m_2 \ge 2$ so that the series (6) converges rectangularly to zero at the point (0,0), while $S_{m_1,1}(0) = \sum_{k=-m_1}^{m_1} k^2 \to \infty$ as $m_1 \to \infty$.

This example shows also that rectangular convergence does not imply that the coefficients tend to zero (or even their boundedness).

B) The *quadratic partial sum* (or *cubical* if $N \ge 3$) $S_k(x)$ is defined by a positive integer k and takes the form

$$S_k(x) = \sum_{|n_1|\le k} \cdots \sum_{|n_N|\le k} c_n e^{inx}. \tag{7}$$

If $\lim_{k\to\infty} S_k(x)$ exists we say that the series (1) is *quadratically* (or *cubically*) *convergent*.

It is clear that the sum (7) is gotten as a special case of the rectangular sum (5) for $m = (k, k, \dots, k)$. Therefore rectangular convergence implies quadratic convergence. The converse is not true, as is easily seen at the hand of the double series

$$\sum_{n_1,n_2=-\infty}^{\infty} c_{n_1 n_2} e^{i(n_1 x_1 + n_2 x_2)}$$

with coefficients $c_{n_1 n_2} = n_1^2 \delta_{n_2 0} - n_2^2 \delta_{n_1 0}$, where again δ_{jk} is the Kronecker symbol. Obviously, this series is quadratically convergent at the point (0,0), but not rectangularly convergent.

C) The *circular partial sum* (or *spherical* if $N \geq 3$) $\tilde{S}_R(x)$ takes the form

$$\tilde{S}_R(x) = \sum_{|n| \leq R} c_n e^{inx}. \tag{8}$$

The series (1) is *circularly convergent* (or *spherically* if $N \geq 3$) if $\lim_{R\to\infty} \tilde{S}_R(x)$ exists.

Rectangular convergence does not imply spherical convergence. As an example we may take the above series (6). Indeed, this series is rectangularly convergent at the point (0,0), but for integers $l > 2$ we have

$$\tilde{S}_l(0) = \sum_{n_1^2 + n_2^2 \leq l^2} c_{n_1 n_2} = 2l^2 \to \infty.$$

Also the converse is not true, i. e. circular convergence does not imply quadratic convergence (and a fortiori not rectangular convergence). As an example we may use a double series (1) with

$$c_{5\cdot 2^k, 0} = -c_{4\cdot 2^k, 3\cdot 2^k} = k, \quad k = 1, 2 \dots,$$

all remaining coefficients being zero. It is clear that $\tilde{S}_R(0) = 0$ for all $R > 0$. However, the sequence of quadratic partial sums $S_{4\cdot 2^k} = -k$ tends to $-\infty$ as $k \to \infty$.

All three partial sums A)–C), clearly, coincide with (4) if $N = 1$. Let us remark that any of these sums may be written in the form

$$S_\omega(x) = \sum_{n \in \omega} c_n e^{inx}, \tag{9}$$

where ω is some finite subset of the lattice $\mathbb{Z}^N$. Now, even if in the case of spherical and cubical sums the subsets ω form a family Ω of enclosing sets, this is not so in the case of rectangular partial sums. In fact, for an acceptable definition it is not necessary to require that the sets ω enclose each other in order to have among them a subsequence of expanding sets exhausting $\mathbb{Z}^N$. These considerations lead us to the following general definitions.

D) Let $\Omega = \{\omega\}$ a family of finite subsets of $\mathbb{Z}^N$ having the following properties:

1) $\forall \omega', \omega'' \in \Omega \quad \exists \omega \in \Omega : \omega' \cup \omega'' \subset \omega$;
2) $\bigcup_{\omega \in \Omega} \omega = \mathbb{Z}^N$.

We define the *partial Ω-sum of the series* (1) by formula (9). We say that the *series* (1) *Ω-converges* at the point x to $f(x)$ if $\forall \varepsilon > 0 \quad \exists \omega_\varepsilon \in \Omega$ such that

$$\forall \omega \in \Omega \, (\omega_\varepsilon \subset \omega) \Rightarrow |S_\omega(x) - f(x)| < \varepsilon.$$

It is clear that if we take for Ω the family of rectangles (parallelotopes), squares (cubes) or disks (balls), then we obtain the rectangular, quadratic and circular partial sums, respectively.

If, on the other hand, the family Ω consist of all finite subsets of $\mathbb{Z}^N$, then only the absolutely convergent sums are Ω-summable.

E) Let Q be any bounded subset of $\mathbb{R}^N$ containing the origin and let Ω consist of all sets of the form $\{n \in \mathbb{Z}^N : n \in \lambda Q\}$, where $\lambda > 0$. The corresponding partial sums have the form

$$S_\lambda(x, Q) = \sum_{n \in \lambda Q \cap \mathbb{Z}^N} c_n e^{inx}. \tag{10}$$

In this case (1) is Ω-convergent if $\lim_{\lambda \to \infty} S_\lambda(x, Q)$ exists. In particular, if Q is the unit ball or the unit cube, then $S_\lambda(x, Q)$ is, respectively, the cubic or the spherical partial sum.

Note that this convergence does not coincide, for no choice of Q whatsoever, with rectangular convergence.

4. Forms of Convergence. Everything that has been said in Sect. 3 refers, of course, also to any multiple series and not necessarily to the trigonometric series (1). On the other hand, as the series (1), which is the main object of study for us, is a function series, one can introduce various forms of convergence. These forms differ from each other not only in the way one takes the partial sums but also in the way they behave on $\mathbb{T}^N$. The most important forms of convergence are uniform convergence, convergence at a fixed point, a. e. convergence, convergence in the metric of $L_p(\mathbb{T}^N)$.

Let us assume that the series (1) converges in a determined sense to a function $f(x)$:

$$f(x) = \sum_{n \in \mathbb{Z}^N} c_n e^{inx}. \tag{11}$$

So far we have not put any restrictions whatsoever on the coefficients c_n, which may have been arbitrary complex numbers. One of the first problems which arises in the study of orthogonal series (to which also trigonometric series belong) is the problem of determing the coefficients. As in the one-dimensional case, the solution to this problem depends on in which sense the series converges to f. If the type of convergence allows one to integrate the series term by term (for example, in the case of uniform convergence or convergence in L_p), then upon multiplying both members of (11) with e^{-inx} and integrating over $\mathbb{T}^N$ we obtain $c_n = f_n$, where the numbers

$$f_n = (2\pi)^{-N} \int_{\mathbb{T}^N} f(x) e^{-inx} dx \tag{12}$$

are referred to as the *Fourier coefficients* of f. It is clear that the Fourier coefficients exist only if $f \in L_1(\mathbb{T}^N)$, i. e. if f is a function summable over $\mathbb{T}^N$ (with respect to the N-dimensional Lebesgue measure). The Fourier coefficients f_n of a function $f \in L_1(\mathbb{T}^N)$ tend to zero as $|n| \to \infty$ (*Riemann-Lebesgue theorem*).

With each function $f \in L_1(\mathbb{T}^N)$ we can thus associate a multiple trigonometric series

$$f(x) \sim \sum_{n \in \mathbf{Z}^N} f_n e^{inx}, \tag{13}$$

termed the *Fourier series* of f. The sign $\sim$ means that the series has been obtained in a purely formal way without any statements about its convergence. The following important problem arises now: must the Fourier series converge in some sense, and if this is the case, does it converge to the function f?

If the equality (11) is satisfied in such a way that term by term integration is not possible (i. e. in the case of pointwise or a. e. convergence), then the coefficients are in general not uniquely determined by this equality. The uniqueness problem which arises in this way can in somewhat different wording be stated as follows: does it follow from the convergence of the series (1) to zero that all coefficients are zero? In the one dimensional case Riemann's theorem gives an exhaustive answer to this question in the case of pointwise convergence, while the counter-example of Menshov gives a negative answer for a. e. convergence (cf. Bari (1961)). Clearly, Menshov's example remains in force also for $N > 1$. What concerns Riemann's theorem, then so far only an analogue is known for $N = 2$: *Cooke's theorem* (1971) to the effect that if a double trigonometric series converges circularly to zero at each point then all coefficients must vanish. If $N \geq 3$ the problem of uniqueness remains open.

From now on (with the exception of Sects. 1 and 2 in Chapter 4), we will restrict further discussion to Fourier series only and we will not consider general trigonometric series.

5. Eigenfunction Expansions. In the beginning of the nineteenth century a new connection between the theory of multiple Fourier series and the theory of partial differential equations was unvealed, namely in the subdomain of that field which is known as spectral theory. In order to make the presentation as transparent as possible we will begin our discussion with the classical *Laplace operator*

$$\Delta = \sum_{k=1}^{N} \frac{\partial^2}{\partial x_k^2}.$$

The operator Δ is considered in the Hilbert space $L_2(\mathbb{T}^N)$ as an unbounded operator with domain of definition $C^\infty(\mathbb{T}^N)$.[1]

As for any two functions u and v in the domain of definition holds[2,3]

$$(\Delta u, v) = (u, \Delta v),$$

$$(\Delta u, u) = -(\nabla u, \nabla u) \leq 0,$$

the operator $-\Delta$ is symmetric and nonnegative. Consequently, by Friedrichs's theorem (cf. Alimov, Il'in, and Nikishin (1976/77)) it has a nonnegative selfadjoint extension, which we denote by $\hat{A}$. It is not hard to see that this selfadjoint extension is unique and coincides with the closure of $-\Delta$.

The operator $\hat{A}$ has in $L_2(\mathbb{T}^N)$ a complete orthonormal system of eigenfunctions

$$\{(2\pi)^{-N/2}e^{inx}\}, \quad n \in \mathbf{Z}^N,$$

corresponding to the eigenvalues $\{|n|^2\}$, $n \in \mathbf{Z}^N$. Like every selfadjoint operator, the extension $\hat{A}$ has in view of von Neumann's theorem a decomposition of unity $\{E_\lambda\}$ with the aid of which it can be written in the form (cf. Alimov, Il'in, and Nikishin (1976/77))

$$\hat{A} = \int_0^\infty \lambda dE_\lambda.$$

It is easy to check that the operators E_λ have the form

$$E_\lambda f(x) = \sum_{|n|^2<\lambda} f_n e^{inx},$$

where f_n are the Fourier coefficients of the function $f \in L_2(\mathbb{T}^N)$ defined with the help of the identity (12). As we have seen, the family $\{E_\lambda f\}$ called the *spectral expansion* of f, coincides with the spherical partial sums of the Fourier series (13).

Let us now consider instead of the Laplace operator an arbitrary differential operator with constant coefficients

$$A(D) = \sum_{|\alpha|\leq m} a_\alpha D^\alpha, \tag{14}$$

where $\alpha = (\alpha_1, \dots, \alpha_N)$ is a *multi-index*, i. e. an element of $\mathbf{Z}^N$ with nonnegative coordinates, $|\alpha| = \alpha_1 + \cdots + \alpha_N$, $D^\alpha = D_1^{\alpha_1} D_2^{\alpha_2} \dots D_N^{\alpha_N}$, $D_j = \frac{1}{i}\frac{\partial}{\partial x_j}$.

---

[1] $C^\infty(\mathbf{T}^N)$ is the class of infinitely differentiable functions on $\mathbf{T}^N$ which are 2π-periodic in each argument.

[2] (u, v) is the inner product in $L_2(\mathbf{T}^N)$.

[3] $\nabla u = \left(\frac{\partial u}{\partial x_1}, \dots, \frac{\partial u}{\partial x_n}\right)$ is the gradient of u.

We associate with each differential operator (14) its *symbol*, an algebraic polynomial in N variables:

$$A(\xi) = \sum_{|\alpha|\leq m} a_\alpha \xi^\alpha,$$

where $\xi = (\xi_1, \dots, \xi_N)$, $\xi^\alpha = \xi_1^{\alpha_1} \dots \xi_N^{\alpha_N}$.

A differential operator is called *elliptic* if its principal symbol

$$A_0(\xi) = \sum_{|\alpha|=m} a_\alpha \xi^\alpha \tag{15}$$

is positive definite, i. e. if for any $\xi \in \mathbb{R}^N$, $\xi \neq 0$, we have $A_0(\xi) > 0$.

If the coefficients of the operator (14) are real, then $A(D)$ is symmetric, i. e.

$$(Au, v) = (u, Av), \quad u, v \in C^\infty(\mathbb{T}^N).$$

If $A(D)$ is also elliptic then by the well-known Gårding's inequality (cf. Hörmander (1983–85), Vol. III) it is bounded from below

$$(Au, u) \geq c\,(u, u), \quad c \in \mathbb{R}^1.$$

Therefore, by Friedrichs's theorem just mentioned, $A(D)$ has a selfadjoint extension $\hat{A}$ and

$$\hat{A} = \int_c^\infty \lambda dE_\lambda,$$

where $\{E_\lambda\}$ is the corresponding resolution of identity. As in the case of the Laplace operator, the eigenfunctions of $\hat{A}$ are the functions $(2\pi)^{-N/2} e^{inx}$, while the eigenvalues equal $A(n)$.

The spectral expansion of a function $f \in L_2(\mathbb{T}^N)$ takes the form

$$E_\lambda(A) f(x) = \sum_{A(n)<\lambda} f_n e^{inx}. \tag{16}$$

Unfortunately, the corresponding partial sum expansions do not refer to the ones considered in A)–C) and E) in Sect. 3, because the domains

$$\{\xi \in \mathbb{R}^N : A(\xi) < \lambda\}$$

are in general not similar for different values of λ. This is explained by the possible presence of lower order terms in the polynomial (14). The situation improves considerably if we assume that the polynomial (14) is homogeneous, i. e. coincides with its principal symbol (15). In this case, putting

$$Q = \{\xi \in \mathbb{R}^N : A_0(\xi) \leq 1\},$$

we get the expansion $E_\lambda(A_0)f$ which coincides with the partial sums already known to us in E) in Sect. 3:

$$E_\lambda(A_0)f = \tilde{S}_\mu(x, Q), \ \mu^m = \lambda.$$

It might appear that the study of the spectral expansions $E_\lambda(A)$ has a very special character, as it refers only to operators defined on $\mathbb{T}^N$. However, analogous reasonings may be carried out also for the operator $A(D)$ considered on an arbitrary domain $\Omega \subset \mathbb{R}^N$. In this case one has in general infinitely many selfadjoint extensions and the corresponding resolutions of unity E_λ do not have any longer the form (16). However, it is remarkable that the spectral behavior of all the different selfadjoint extensions of the one operator $A(D)$ is basically the same and mimics the one of (16). The mathematical formulations of this not completely clearly formulated statement comprises the union of a great number of theorems known as equiconvergence theorems. The uncontested importance of these theorems is seen by the fact that they allow one to reduce the study of the immense variety of spectral expansions to the study of one object: the multiple Fourier series (or integral). A characteristic feature of equiconvergence theorems is that their validity usually involves a class of functions: for each function f in some class W we have

$$\tilde{E}_\lambda f - E_\lambda f \to 0 \text{ for } \lambda \to \infty.$$

From this one can conclude that if $f \in W$ then the convergence of $\tilde{E}_\lambda f$ holds if and only if $E_\lambda f$ converges.

However, as often is the case, precisely this feature turns out to be a defect. Namely, in many situations equiconvergence is established for precisely those classes one has the convergence of $E_\lambda f$ and, therefore, of $\tilde{E}_\lambda f$. In those classes where $E_\lambda f$ does not converge one does not have as a rule equiconvergence and therefore one cannot say anything about the behavior of $\tilde{E}_\lambda f$. In other words, equiconvergence theorems apply effectively when it is required to prove convergence but they are somewhat harder to apply in divergence proofs.

Another important fact is that the "lower order" coefficients a_α, $|\alpha| < m$, in (14) do not influence the convergence of the spectral expansion $E_\lambda f$ of the operator $A(D)$, provided the function under view is sufficiently smooth. In other words, one has equiconvergence not only for expansions of one and the same operator but also for expansions of different operators A_1 and A_2, again provided only their principal parts coincide. Using this fact one can, in particular, reduce the study of the partial sum (16) to the study of the simpler expansion $E_\lambda(A_0)f$.

6. Summation Methods. As we have already noted in Sect. 4, one of the main problems of harmonic analysis is the reconstruction of functions from their expansion. The definition of the sum of a Fourier series as the limit of partial sums does not always solve this problem. The point is that for not very smooth functions (and exactly such functions have the most interesting and from the practical point of view important expansions) successive strongly oscillating terms of the partial sums have a very big influence, which does not correspond to their unsignificant rôle in the characterization of the function to be reconstructed. As a result the sequence of partial sums does not approach the function but oscillates around it. However, these oscillations have, as a rule, a regular behavior, so taking, for example, arithmetic means of the partial sums we may expect that these means better approximate the function under consideration. Mathematically this is equivalent to taking, in the calculation of the partial sum, each term of the series with a weight which decreases in size as we increase the index.

This heuristic device works in all cases where the partial sums are parametrized by a natural index, for example, for quadratic sums. Let, say,

$$S_k = \sum_{|n_1|\leq k} \sum_{|n_2|\leq k} c_{n_1 n_2} e^{i(n_1 x_1 + n_2 x_2)}$$

be the partial sum of a double series. Then, after some simple transformations, it is not hard to see that

$$\sigma_l = \frac{1}{l} \sum_{k=0}^{l-1} S_k = \sum_{|n_1|\leq l-1} \sum_{|n_2|\leq l-1} \alpha_{n_1 n_2}^{(l)} c_{n_1 n_2} e^{i(n_1 x_1 + n_2 x_2)},$$

where $\alpha_{n_1 n_2}^{(l)} = 1 - \frac{1}{l}\max(|n_1|, |n_2|)$.

In particular, if $n_1 = l - 1$ or $n_2 = l - 1$ then $\alpha_{n_1 n_2}^{(l)} = \frac{1}{l}$. The means σ_l usually behave better than S_k as $l \to \infty$.

We may proceed in an analogous manner when the partial sums are defined by an continuous parameter, for example, in the case of the spectral resolution E_λ. In this case it is natural to define integral means as follows:

$$\sigma_\lambda f = \frac{1}{\lambda} \int_0^\lambda E_t f\, dt.$$

Integrating by parts, the last formula takes the form

$$\sigma_\lambda f = \int_0^\lambda \left(1 - \frac{t}{\lambda}\right) dE_t f.$$

Each time that the partial sums $E_\lambda f$ converge the mean σ_λ will also converge to the same value. The converse is not true. For example, if $E_\lambda f$ oscillates asymptotically as $\cos\lambda$ then σ_λ tends to 0 as $\frac{1}{\lambda}\sin\lambda$. This leads to the hope that one might be able to sum the series with the aid of the means σ_λ in cases when the partial sums do not have a limit.

It is natural to expect that the chances for succcess increase if we integrate E_λ several times. After an s-fold integration of $E_t f$ over the interval $[0, \lambda]$

and a subsequent normalization we arrive after some transformations at the quantities

$$E_\lambda^s f = \int_0^\lambda \left(1 - \frac{t}{\lambda}\right)^s dE_t f, \tag{17}$$

which are known as the *Riesz means* of order s. We say that the spectral expansion (16) is *summable* to $f(x)$ by the *Riesz method* of order s if

$$\lim_{\lambda \to \infty} E_\lambda^s f(x) = f(x).$$

Let us note that the integral in (17) makes sense for any real $s \geq 0$ and even for complex s with $\operatorname{Re} s \geq 0$, thanks to which one can use interpolation theorems in the study of Riesz means.

If A is an elliptic operator then $E_\lambda f$ is given by (16). Inserting this into (17) gives

$$E_\lambda^s f(x) = \sum_{A(n)<\lambda} \left(1 - \frac{A(n)}{\lambda}\right)^s f_n e^{inx}. \tag{18}$$

In the case of the Laplace operator the Riesz means take an especially simple form:

$$\sigma_\lambda^s f(x) = \sum_{|n|^2<\lambda} \left(1 - \frac{|n|^2}{\lambda}\right)^s f_n e^{inx}. \tag{19}$$

These *Riesz-Bochner means* for spherical partial sums were first studied in detail by Bochner (1936).

The Riesz means may be viewed as a regularization of the partial sums, and in many cases their asymptotic behavior gets better when $\operatorname{Re} s$ is increased.

The regular summation methods in the case of partial sums defined by several parameters, for example rectangular partial sums, are somewhat more difficult to deal with. In this case one can likewise define means (cf. Zygmund (1968), Vol. II, Chapter XVII).

7. Multiple Fourier Integrals. Exactly as in the classical (one dimensional) case trigonometric series are intimately connected with Fourier integrals, likewise multiple Fourier series admit a continuous analogue – the *multiple Fourier integrals.*

If $f \in L_1(\mathbb{R}^N)$ then its *Fourier transform* is defined by the formula

$$\hat{f}(\xi) = (2\pi)^{-N/2} \int_{\mathbf{R}^N} f(x) e^{-ix\xi} dx. \tag{20}$$

The function can be recovered in a unique way from its Fourier transform by the formula

$$f(x) = (2\pi)^{-N/2} \int_{\mathbf{R}^N} \hat{f}(\xi) e^{ix\xi} d\xi. \tag{21}$$

However, in contrast to (20) this formula does not make sense for all $f \in L_1(\mathbb{R}^N)$.

The formal requirement of the convergence of the improper integral (21), equivalent to the condition $\hat{f} \in L_1(\mathbb{R}^N)$, restricts considerably the class of functions f which can be expanded in a multiple Fourier integral. Therefore, there arises the question of the special convergence of the improper integral (21), analogous to the convergence problem for multiple trigonometric series.

As in Sect. 3 D), let us consider a family of bounded sets $\omega \subset \mathbb{R}^N$ enjoying the following properties:

1) $\forall \omega_1, \omega_2 \in \Omega \quad \exists \omega \in \Omega : \omega_1 \cup \omega_2 \subset \omega$;

2) $\bigcup_{\omega \in \Omega} \omega = \mathbb{R}^N$.

We define the corresponding partial integral by the formula

$$I_\omega(x, f) = (2\pi)^{-N/2} \int_\omega \hat{f}(\xi) e^{ix\xi} d\xi. \tag{22}$$

We say that the integral (21) *converges to $f(x)$ with respect to the family* Ω if

$$\forall \varepsilon > 0 \quad \exists \omega_\varepsilon \subset \Omega \quad \forall \omega \subset \Omega : \omega_\varepsilon \subset \omega \Rightarrow |I_\omega(x, f) - f(x)| < \varepsilon.$$

It is clear that this includes *rectangular, quadratic (cubic)* and *circular (spherical) convergence of the integral* (21).

If $\hat{A}$ is a selfadjoint elliptic operator in $L_2(\mathbb{R}^N)$ generated by the differential expression (14) and if E_λ is the corresponding spectral resolution, then the spectral expansion of any element $f \in L_2(\mathbb{R}^N)$ has the form

$$E_\lambda f(x) = (2\pi)^{-N/2} \int_{A(\xi) < \lambda} \hat{f}(\xi) e^{ix\xi} d\xi. \tag{23}$$

This formula needs some explanations, because we have defined $\hat{f}$ only for $f \in L_1(\mathbb{R}^N)$. If $f \in L_2(\mathbb{R}^N)$ then the integral (20) need not exist in Lebesgue's sense so there arises the question of a workable definition of $\hat{f}$. The key to this problem is provided by *Plancherel's theorem* according to which for each function $f \in L_1(\mathbb{R}^N) \cap L_2(\mathbb{R}^N)$ we have *Parseval's formula*

$$\|\hat{f}\|_{L^2(\mathbb{R}^N)} = \|f\|_{L^2(\mathbb{R}^N)}.$$

This means that the Fourier transform is an isometry and so, being defined on the dense subset $L_1(\mathbb{R}^N) \cap L_2(\mathbb{R}^N)$ of the Hilbert space $L_2(\mathbb{R}^N)$, it admits a unique extension to the whole of $L_2(\mathbb{R}^N)$ by passing to the closure. This extension assigns to each function $f \in L_2(\mathbb{R}^N)$ its *Fourier-Plancherel transform*, which coincides with the usual Fourier transform when f is summable.[4]

[4] If f is locally summable then $\hat{f}$ can be defined in the sense of the theory of distributions. If $f \in L_1$ or $f \in L_2$ then this definition coincides with the previous one.

The process of closure can be performed in several concrete ways, i. e. with the aid of different sequences $f_n \in L_1(\mathbb{R}^N) \cap L_2(\mathbb{R}^N)$. The simplest is to put

$$f_n(x) = \chi_n(x) f(x),$$

where $\chi_n(x)$ is the characteristic function of expanding sets $\omega_n \subset \mathbb{R}^N$ which exhaust $\mathbb{R}^N$. Then, irrespective of the choice of ω_n, the sequence (cf. (22))

$$\hat{f}_n(\xi) = (2\pi)^{-N/2} \int_{\omega_n} f(x) e^{ix\xi} dx$$

converges in the metric of $L_2(\mathbb{R}^N)$ to a function $\hat{f}(\xi)$ which we take as the Fourier transform of f. Usually one takes for ω_n concentric balls or cubes.

For the spectral expansion (23) we can, as in the case of series, define Riesz means of order $s \geq 0$:

$$E_\lambda^s f(x) = (2\pi)^{-N/2} \int_{A(\xi) \leq \lambda} \left(1 - \frac{A(\xi)}{\lambda}\right)^s \hat{f}(\xi) e^{ix\xi} d\xi. \tag{24}$$

In Sect. 5, speaking of equiconvergence theorems, we noted that the spectral expansions $E_\lambda f$, corresponding to different selfadjoint expansions of one and the same operator $A(D)$, behave on the whole in a unique way and so make it sufficient to study one of these expansions in order to be able to make conclusions about behavior of all the others. Thus, the most convenient object for a detailed investigation are the expansions of the form (23) or their Riesz means (24). In the following section we shall try to elaborate this assertion.

8. The Kernel of the Fourier Integral Expansion. The partial integral $I_\omega(x, f)$ defined in (22) may be transformed writing instead of $\hat{f}$ the integral to the right in (20) and then changing the order of integration. This yields the formula

$$I_\omega(x, f) = \int_{\mathbb{R}^N} K(x - y, \omega) f(y) dy, \tag{25}$$

with

$$K(x, \omega) = (2\pi)^{-N} \int_\omega e^{ix\xi} d\xi. \tag{26}$$

Thus, every partial Fourier integral is an integral operator whose kernel depends on the difference. It is a remarkable fact that the kernel (26) is the Fourier transform of the characteristic function $\chi(x)$ of the set ω.

Let us assume that the set ω has the form

$$\omega = \mu Q \equiv \{\xi \in \mathbb{R}^N : \frac{1}{\mu}\xi \in Q\},$$

where $\mu > 0$ and Q is a suitable bounded neighborhood of the origin. Then

$$K(x, \mu Q) = (2\pi)^{-N} \int_{\mu Q} e^{ix\xi} d\xi = (2\pi)^{-N} \mu^N \int_Q e^{i\mu x\xi} d\xi,$$

i. e.

$$K(x, \mu Q) = \mu^N K(\mu x, Q).$$

We see that one can get complete information about the kernel $K(x, \mu Q)$ by studying the integral

$$K(x, Q) = (2\pi)^{-N} \int_Q e^{ix\xi} d\xi \tag{27}$$

as a function of x. This function is continuous and further analytic; it is clear that it is subject to the estimate

$$|K(x, Q)| \le |K(0, Q)| = (2\pi)^{-N} |Q|.$$

At infinity $K(x, Q)$ tends to zero oscillating. The way it tends to zero depends on the geometry of the set Q and determines, in turn, the behavior of the partial sums of the Fourier integrals.

If $A(D)$ is a homogenous elliptic polynomial of order m, i. e. $A(t\xi) = t^m A(\xi)$, then the integral (23) belongs to the previous class with the domain Q of the form

$$Q = \{\xi \in \mathbb{R}^N : A(\xi) \le 1\}. \tag{28}$$

Thus, the behavior of the spectral expansion corresponding to the operator $A(D)$ is closely connected with the geometry of the level lines of the polynomial $A(\xi)$. This remarkable observation has stimulated the study of Fourier transforms of characteristic functions of various sets. One of the most powerful methods in the study of oscillatory integrals is the *method of stationary phase.* The elaboration of this method has led to the discovery of the following rule: the "more convex" the set Q is, the faster the function $K(x, Q)$ decreases as $|x| \to \infty$. The most convex set is the ball, and therefore the function

$$K(x) = (2\pi)^{-N} \int_{|\xi|<1} e^{ix\xi} d\xi \tag{29}$$

has optimal behavior.

The integral in the right hand side of (29) (Titchmarsh (1958)) is computed by passing to a system of spherical coordinates and turns out to be

$$K(x) = (2\pi)^{-N} |x|^{-\frac{N}{2}} J_{\frac{N}{2}}(|x|), \tag{30}$$

where $J_\nu(t)$ is the Bessel function of first kind and order ν. As

$$|J_\nu(t)| \le \frac{\text{const}}{\sqrt{t}}, \quad t > 0, \tag{31}$$

it follows from (30) that

$$|K(x)| \leq \text{const} \cdot |x|^{-\frac{N+1}{2}}. \tag{32}$$

This is the best estimate for functions of the type (27), which is uniform in all directions.

Of course, in special directions, depending on the geometry of Q, one may improve the estimate (32). Let, for example, Q be the cube

$$Q = \{\xi \in \mathbb{R}^N : |\xi_j| \leq 1\}.$$

Then clearly

$$K(x, Q) = \pi^{-N} \prod_{k=1}^{N} \frac{\sin x_k}{x_k}.$$

If $x \to \infty$ then along the diagonal $x_1 = x_2 = \cdots = x_N$

$$|K(x)| \leq \frac{\text{const}}{|x|^N},$$

which, apparently, is better than (32), but if $x \to \infty$ along a coordinate axis then the estimate

$$|K(x)| \leq \frac{\text{const}}{|x|}, \tag{33}$$

is sharp, i. e. the kernel oscillates rather slowly. Of course, one can also construct more exotic examples of sets Q for which the kernel $K(x, Q)$ oscillates even slower than (33), but for a large class of sets of type (28) the uniform estimates are included between (32) and (33).

Useful objects in this study are the elliptic operators of the special type

$$A_m(D) = \sum_{k=1}^{N} D_k^m, \; m = 2m', \tag{34}$$

which were first considered in detail for this very purpose by Peetre (1964). The sets Q_m corresponding to these operators have the form

$$Q_m = \{\xi \in \mathbb{R}^N : \xi_1^m + \cdots + \xi_N^m \leq 1\}. \tag{35}$$

If $m = 2$ the set Q_m is the N-dimensional ball and therefore the spectral expansion (23) coincides with the spherical partial integrals. If m increases the boundary flattens, and as $m \to \infty$ the set approaches a cube and the spectral expansion (23) goes over into the cubical partial integrals. These examples convince us that the order of decrease of the Fourier transform of the characteristic function of a set Q depends on the convexity of its boundary

∂Q. As a numerical measure for the convexity of ∂Q we may take the number of the non-zero principal curvatures at each point of ∂Q (cf. Ashurov (1983a)).

Indeed, if the boundary ∂Q of the domain (28) has r $(0 \le r \le N-1)$ non-zero principal curvatures then the Fourier transform (27) satisfies the estimate

$$|K(x,Q)| \le \text{const}\,|x|^{-\frac{1}{2}(r+2)}. \tag{36}$$

Here we assume that Q is convex.

Let us recall the simplest definition of *principal curvatures* of an $(N-1)$-dimensional hypersurface M at the point $x \in M$. Let T_xM be the tangent plane and let $y_N = f(y_1, \dots, y_{N-1})$ be the equation of M in a neighborhood of x, $(y_1, y_2, \dots, y_{N-1}) \in T_xM$. Then the eigenvalues of the matrix $\left\{\dfrac{\partial^2 f(x)}{\partial y_j \partial y_k}\right\}$ are called the principal curvatures of M. Note that the principal curvatures are invariants of M, i. e. they are independent of the coordinate system considered.

One can obtain a more accurate measure of $K(x,Q)$ if one takes into account the order with which the principal curvatures vanish, which however constitutes a much harder issue.

9. The Dirichlet Kernel. The discussion in the previous section suggests to study in an analogous way the behavior of partial sums of multiple Fourier series. Let us turn anew to Example D) in Sect. 3:

$$S_\omega(x,f) = \sum_{n\in\omega} f_n e^{inx}.$$

If we replace the coefficients f_n in this sum by their values given by formula (12), we obtain

$$S_\omega(x,f) = \int_{\mathbf{T}^N} D_\omega(x-y) f(y)dy, \tag{37}$$

where

$$D_\omega(x) = (2\pi)^{-N} \sum_{n\in\omega} e^{inx}. \tag{38}$$

Thus, each set $\omega \in \mathbb{Z}^N$ defines an integral operator with a kernel which depends only on the difference. The behavior of kernels of the form (38) can be studied by comparing them with the functions (26) (the proofs of most of the equiconvergence theorems are based on this fact). For instance, in the case of spherical partial sums it is, taking into account (30), natural to expect that

$$\sum_{|n|\le\mu} e^{inx} \sim (2\pi)^{-N/2} \mu^{\frac{N}{2}} \frac{J_{\frac{N}{2}}(\mu|x|)}{|x|^{\frac{N}{2}}}. \tag{39}$$

The sign $\sim$ means that the ratio these two quantities in some sense behaves in the same way for large values of μ. In order to give this assertion an exact meaning we must estimate their difference, which is a far from simple issue.

Let us note that the comparison of the kernels (38) and (26) has an interest of its own, independent of the problem of convergence of partial sums of Fourier series. For example, the comparison (39) is extremely important in number theory even for $x = 0$:

$$\sum_{|n|\le\mu} 1 \sim \frac{\pi^{\frac{N}{2}}}{\Gamma(\frac{N}{2}+1)}\mu^N. \tag{40}$$

To the left stands the number of points of $\mathbb{Z}^N$ contained inside a sphere of radius μ. This can also be interpreted as $\sum_{k<\mu^2} \nu_k(N)$, where $\nu_k(N)$ is the number of solutions of the Diophantine equation

$$n_1^2 + n_2^2 + \cdots + n_N^2 = k.$$

In the case $N = 2$ estimating the difference of the quantities in (40) constitutes the so-called *circle problem* in number theory.

When comparing the kernels (26) and (38) we should bear in mind that the corresponding integral operators act on different function spaces: on the space $L_1(\mathbb{R}^N)$ of summable functions on $\mathbb{R}^N$ and on the space $L_1(\mathbb{T}^N)$ of functions on $\mathbb{R}^N$ which are 2π-periodic in each argument and summable on $\mathbb{T}^N$. It is clear that these classes are different and that their intersection consists of the zero function only.

Let us assume that $f \in L_1(\mathbb{T}^N)$ vanishes near the boundary of the cube $\mathbb{T}^N$. Then the function

$$g(x) = \begin{cases} f(x), & x \in \mathbb{T}^N, \\ 0, & x \notin \mathbb{T}^N \end{cases}$$

is in $L_1(\mathbb{R}^N)$ and preserves all properties of f in the interior of $\mathbb{T}^N$. Conversely, if $g \in L_1(\mathbb{R}^N)$ has its support in the interior of $\mathbb{T}^N$, then if we shift its graph along the coordinate axes with steps which are multiples of 2π, we get a periodic function $f \in L_1(\mathbb{T}^N)$ which coincides with g on $\mathbb{T}^N$, i. e.

$$f(x) = \sum_{n\in\mathbf{Z}^N} g(x + 2\pi n). \tag{41}$$

Clearly, the Fourier coefficients f_n of f coincide up to a factor with the Fourier transform $\hat{g}$ of g taken at the point n:

$$f_n = (2\pi)^{-N}\int_{\mathbf{T}^N} f(x)e^{-inx}dx = (2\pi)^{-N}\int_{\mathbf{R}^N} g(x)e^{-inx}dx = (2\pi)^{-N/2}\hat{g}(n).$$

In this case we have

$$f(x) = \sum_{n\in\mathbf{Z}^N} f_n e^{inx} = (2\pi)^{-N/2}\sum_{n\in\mathbf{Z}^N} \hat{g}(n)e^{inx},$$

so comparing this with (41) we obtain

$$\sum_{n\in\mathbf{Z}^N} g(x+2\pi n) = (2\pi)^{-N/2} \sum_{n\in\mathbf{Z}^N} \hat{g}(n)e^{inx}. \tag{42}$$

The equality (42) is called *Poisson's summation formula.* The preceding reasoning has a heuristic character and cannot serve for the foundation of (42). However, if the series to the left and to the right are uniformly convergent for $x \in \mathbb{T}^N$, it is not hard to show the validity of the formula. Then it is not necessary to assume that the function $g \in L_1(\mathbb{T}^N)$ vanishes off $\mathbb{T}^N$. For instance, if g is such that it and its Fourier transform are subject to the estimates

$$\begin{aligned} |g(x)| &\le \text{const}\,(1+|x|)^{-N-\varepsilon}, \\ |\hat{g}(\xi)| &\le \text{const}\,(1+|\xi|)^{-N-\varepsilon}, \quad \varepsilon > 0, \end{aligned} \tag{43}$$

then the series converge absolutely and uniformly and (42) is in force.

The Poisson summation formula is very effective in the solution of various problems in harmonic analysis, for instance, in the summation of multiple Fourier series. As an illustration consider

$$K(x) = \sum_{n\in\mathbf{Z}^N} \varphi(x+2\pi n), \tag{44}$$

where φ satisfies estimates of the type (43). Then by Poisson's summation formula we get

$$K(x-y) = (2\pi)^{-N/2} \sum_{n\in\mathbf{Z}^N} \hat{\varphi}(n)e^{in(x-y)}.$$

Upon multiplying both sides of this equality by $f \in L_1(\mathbb{T}^N)$ and integrating over $\mathbb{T}^N$ we obtain:

$$K * f(x) = (2\pi)^{-N/2} \sum_{n\in\mathbf{Z}^N} \hat{\varphi}(n) f_n e^{inx}, \tag{45}$$

with the *convolution* $K * f$ defined in the well-known manner:

$$K * f(x) = \int_{\mathbb{T}^N} K(x-y)f(y)dy.$$

If $\varphi(x)$ is the characteristic function of a set ω, then formula (45) gives an expression for the corresponding partial sums of the Fourier series of f. Unfortunately, the Fourier transform of characteristic functions of bounded sets does not oscillate sufficiently fast in order to allow us to apply the Poisson summation formula. It is a different thing if we "smoothen" the function φ, i. e. if we pass from the partial sums to their means.

Consider for example the function

$$\hat{\varphi}_\mu(\xi) = \begin{cases} \left(1 - \frac{|\xi|^2}{\mu^2}\right)^s, & \text{if } |\xi| \leq \mu, \\ 0, & \text{if } |\xi| > \mu. \end{cases} \tag{46}$$

Then

$$\varphi_\mu(x) = (2\pi)^{-N/2} \int_{|\xi|<\mu} \left(1 - \frac{|\xi|^2}{\mu^2}\right)^s e^{ix\xi} d\xi.$$

The integral is computed by passing to a spherical coordinate system and equals

$$\varphi_\mu(x) = (2\pi)^{-N/2} 2^s \Gamma(s+1) \mu^N \frac{J_{\frac{N}{2}+s}(|x|\mu)}{(|x|\mu)^{\frac{N}{2}+s}}. \tag{47}$$

Using an estimate for the Bessel function it follows that

$$|\varphi_\mu(x)| \leq \text{const} \cdot |x|^{-\frac{N+1}{2}-s},$$

so if $s > \frac{N-1}{2}$ then φ_μ satisfies an estimate of type (43). Therefore (45) is applicable to φ_μ, for which it takes the form

$$\sum_{|n|\leq\mu} \left(1 - \frac{|n|^2}{\mu^2}\right)^s f_n e^{inx} = K_\mu * f(x), \tag{48}$$

with

$$K_\mu(x) = c_N \mu^{-\frac{N}{2}-s} \sum_{n\in\mathbf{Z}^N} |x + 2\pi n|^{-\frac{N}{2}-s} J_{\frac{N}{2}+s}(\mu|x + 2\pi n|). \tag{49}$$

The formulae (48) and (49) give a representation of the Riesz means of spherical partial sums for $s > \frac{N-1}{2}$. The order $s = \frac{N-1}{2}$ was termed the *critical order* by Bochner, because many properties of the Riesz means for higher orders fail for $s = \frac{N-1}{2}$. The behavior gets even worse if $s < \frac{N-1}{2}$. In this case for uniform convergence, say, one has to invoke classes of smooth functions, about which we will say a few words in the following section.

10. Classes of Differentiable Functions. The classes of differentiable functions in several variables which have been studied most are the *Sobolev classes* $W_p^l(G)$ (Sobolev (1950)), where $1 \leq p < \infty$, $l = 1, 2, \ldots$ and $G \subset \mathbb{R}^N$.

A function $f \in L_p(G)$ belongs to $W_p^l(G)$ if all partial derivatives $D^\alpha f$ (in the sense of distributions) of order $|\alpha| = l$ are in $L_p(G)$, i. e. if the norm

$$\|f\|_{W_p^l} = \|f\|_{L_p} + \sum_{|\alpha|=l} \|D^\alpha f\|_{L_p} \tag{50}$$

is finite.

If $N = 1$ membership of f in W_p^l means that f has $l-1$ continuous derivatives and that $f^{(l-1)}$ is absolutely continuous with $f^{(l)}$ belonging to L_p. Simple examples reveal that for higher dimensions $f \in W_p^l$, in general, does not even imply that f is continuous. However, if p and l are sufficiently large then each function in W_p^l is continuous, i. e. the space $W_p^l(G)$ is continuously imbedded in $C(G)$. Namely, if $pl > N$ then the identity operator

$$I : W_p^l \to C$$

is continuous (and even compact). This statement is the content of one of the famous *Sobolev imbedding theorems* (Sobolev (1950)).

The classes $W_p^l(\mathbb{R}^N)$ can also be described in terms of the Fourier transform. Namely, $f \in W_p^l(\mathbb{R}^N)$ if and only if there is a (uniquely determined) function $g \in L_p(\mathbb{R}^N)$ such that

$$\hat{f}(\xi) = (1 + |\xi|^2)^{-l/2} \hat{g}(\xi). \tag{51}$$

An equivalent norm in $W_p^l(\mathbb{R}^N)$ is provided by the quantity

$$\|f\|_{\widetilde{W}_p^l} = \|g\|_{L_p}, \quad p \geq 1. \tag{52}$$

That the norms (50) and (52) are equivalent can easily be checked for $p = 2$. Indeed, if $p = 2$ then (50) is equivalent to the norm

$$\|f\|_l^2 = \|f\|_{L_2}^2 + \sum_{|\alpha|=l} \|D^\alpha f\|_{L_2}^2.$$

As $(\widehat{D^\alpha f})(\xi) = \xi^\alpha \hat{f}(\xi)$ it follows from Parseval's formula that

$$\|f\|_l = \left\| \left(1 + \sum_{|\alpha|=l} |\xi^\alpha|^2 \right)^{1/2} \hat{f}(\xi) \right\|_{L_2},$$

while the norm in the right hand side is equivalent to the norm (51)–(52), as clearly

$$c_1(1 + |\xi|^2)^l \leq 1 + \sum_{|\alpha|=l} |\xi^\alpha|^2 \leq c_2(1 + |\xi|^2)^l.$$

If $p \neq 2$ the proof of the equivalence is considerably harder.

In order to simplify the notation it is expedient to introduce the direct and inverse Fourier transforms:

$$\mathcal{F}f(\xi) = (2\pi)^{-N/2} \int_{\mathbb{R}^N} f(x) e^{-ix\xi} dx, \tag{53}$$

$$\mathcal{F}^{-1}f(x) = (2\pi)^{-N/2} \int_{\mathbb{R}^N} f(\xi) e^{ix\xi} d\xi. \tag{54}$$

Then the definition of the norm (51)–(52) can also be written as

$$\|f\|_{W_p^l} = \|\mathcal{F}^{-1}(1+|\xi|^2)^{l/2}\mathcal{F}f\|_{L_p}, \quad l = 1, 2, \ldots. \tag{55}$$

Let us remark that the exponent l in (50) has to be a positive integer, because it is not completely clear which sense has to be given to the norm for nonintegral l. The situation improves if we look at the equivalent norm (55). In this definition we can take for l any positive number. The class of functions $f \in L_p(\mathbb{R}^N)$ which for a given fixed number $l > 0$ make the norm (55) finite is termed the *Liouville class*[5] $L_p^l(\mathbb{R}^N)$. Thus

$$\|f\|_{L_p^l} = \|\mathcal{F}^{-1}(1+|\xi|^2)^{l/2}\mathcal{F}f\|_{L_p}, \quad l > 0. \tag{56}$$

The classes L_p^l coincide for l integer with the Sobolev classes W_p^l and constitute a natural continuation of the latter to noninteger values of l. Besides, they are natural precisely from the point of view of harmonic analysis. If we however want to introduce fractional derivatives, having in mind the exact description of the traces of functions in Sobolev classes to subspaces of lower dimension (and exactly these questions are the most appropriate in the theory of boundary problems for partial differential equations, for whose solution also the classes W_p^l were first introduced), then the most natural classes are the *Sobolev-Slobodetskiĭ classes*.

Another approach to the problem of function spaces with a fractional smoothness exponent has been developed in approximation theory. Here the *Nikol'skiĭ classes* $H_p^l(\mathbb{T}^N)$ are the most natural. They consist of those functions which admit an approximation in the $L_p(\mathbb{T}^N)$-metric by trigonometric polynomials of degree n[6] with remainder $O(n^{-l})$. Finally, a very general approach to this problem is provided by the *Besov classes* $B_{p\theta}^l$ with an auxiliary parameter θ: if $p = \theta$ these are the *Sobolev-Slobodeckiĭ classes*, if $\theta = \infty$ the Nikol'skiĭ classes and if $p = \theta = 2$ the Liouville classes L_2^l (Nikol'skiĭ (1977)).

In order not to complicate the presentation we will in the sequel limit ourselves to the Liouville classes L_p^l, justifying this choice by the fact that all the above classes approximately coincide with L_p^l (Nikol'skiĭ (1977)):

$$A_p^{l+\varepsilon} \subset L_p^l \subset A_p^{l-\varepsilon}, \; \varepsilon > 0,$$

where A_p^l denotes any of the classes of Sobolev-Slobodeckiĭ, Nikol'skiĭ or Besov (the latter with arbitrary θ).

---

[5] *Translator's Note.* Also known as *Bessel potential class.*

[6] That is, sums of the form $\sum_{|k|\le n} a_k e^{ikx}$.

If $f \in L_p^l$ then clearly f belongs to a Liouville class for any smaller l, but, even more important, then the summation exponent p can be improved. Let us illustrate this at the hand of the function

$$f(x) = |x|^\varepsilon, \quad \varepsilon > 0, x \in \mathbb{T}^N.$$

It is not hard to see that for all α with $|\alpha| = l$ the derivative $D^\alpha f$ has the singularity $|x|^{\varepsilon - l}$. Therefore $D^\alpha f \in L_p$, i. e. $f \in W_p^l$, if $|x|^{\varepsilon - l} \in L_p$, i. e. if $l - N/p < \varepsilon$.

Consequently, the function under consideration belongs to a whole family of spaces W_p^l for which the quantity $l - N/p$ has one and the same value. This quantity characterizes the integrability of the top derivatives in (50), and l has to do with the increase of the singularity which arises when differentiating, while N/p is the decrease coming from the contribution (the larger the higher is the dimension) introduced in the integral from the Jacobian.

These elementary heuristic considerations, arising by considering the simplest functions of a rather special type, have nevertheless a very general character. It turns out that every Liouville class L_p^l is imbedded in another such class $L_{p_1}^{l_1}$ with $l_1 < l$, provided only that the quantity $l - N/p$ is the same (cf. Nikol'skiĭ (1977)). In other words, we have the imbedding $L_p^l \to L_{p_1}^{l_1}$ provided

$$l - \frac{N}{p} = l_1 - \frac{N}{p_1}, \quad l > l_1,$$

i. e.

$$l_1 = l - N\left(\frac{1}{p} - \frac{1}{p_1}\right), \quad p < p_1. \tag{57}$$

The periodic analogue $L_p^l(\mathbb{T}^N)$ of $L_p^l(\mathbb{R}^N)$ is obtained if we replace the direct Fourier transform in (56) by the Fourier coefficient and the inverse transform by the series:

$$\|f\|_{L_p^l} = \left\| \sum_{n \in \mathbf{Z}^N} (1 + |n|^2)^{l/2} f_n e^{inx} \right\|_{L_p}. \tag{58}$$

The local properties of functions in $L_p^l(\mathbb{T}^N)$ and $L_p^l(\mathbb{R}^N)$ are the same: if $\eta \in C_0^\infty(\mathbb{T}^N)$ then the sets $\{f\eta : f \in L_p^l(\mathbb{T}^N)\}$ and $\{f\eta : f \in L_p^l(\mathbb{R}^N)\}$ coincide.

Besides the spaces L_p^l we require also the classical *Hölder spaces* C^l, which for all positive integers l consist of all functions f having continuous partial derivatives of all orders α with $|\alpha| \le l$. For fractional $l > 0$ the classes C^l are usually defined as follows: if $l = m + \kappa$, where $m = 0, 1, \dots, 0 < \kappa < 1$, then $f \in C^l$ if and only if

$$|D^\alpha f(x) - D^\alpha f(y)| \le \text{const} \cdot |x - y|^\kappa$$

for every α with $|\alpha| = l$. As usual, $C^l(\mathbb{T}^N)$ denotes the class of 2π-periodic functions satisfying the above conditions.

The function $f(x) = |x|^\varepsilon$, $0 < \varepsilon < 1$, is a typical representative of a function of class C^l. We remarked already that this function belongs to every Liouville class L_p^l with $l - N/p < \varepsilon$. This fact leads to the idea that there must be a connection between the Liouville and Hölder classes. Indeed, such a connection is provided by the following *imbedding theorem* (Nikol'skiĭ (1977)):

$$L_p^l \to C^\varepsilon \quad \text{where} \quad e = l - \frac{N}{p} > 0. \tag{59}$$

The meaning of this theorem is that the existence of derivatives of high order which are summable to a large power guarantees for a function its smoothness in the classical sense.

11. A Few Words About Further Developments. First we say a few words about convergence of multiple Fourier series and integrals in the most important metrics – the uniform metric and the L_p metric. Then we shall discuss the problem of a. e. convergence, which is one of the most attractive problems in the metric theory of functions. The problem of L_p convergence is closely connected with the theory of multipliers, the development of which has been one of the decisive motives for the introductions of pseudo-differential operators, and then Fourier integral operators.

Many important results, which are technically hard to formulate, fall outside the framework of this exposition. A desire to adhere to the style dictated by the intentions of this Encyclopaedia, attempting to provide a general first introduction to the subject, has forced us to put less emphasis on questions of priority, rather concentrating on the highlights of the, in our opinion, most important developments in the theory of multiple Fourier series.

To a reader who wishes to penetrate deeper into the theory of multiple series we recommend the books Stein and Weiss (1971), Chapter VIII and Zygmund (1968), Vol. II, Chapter XVIII, and further the surveys Alimov, Il'in, and Nikishin (1976/77) and Zhizhiashvili (1973).

Chapter 1
Localization and Uniform Convergence

§1. The Localization Principle

1.1. On the Localization Problem. The classical *Riemann localization theorem* states that the convergence or divergence of a "one-dimensional" Fourier series at a given point depends only on the behavior of the function $f \in L_1$

in an arbitrary small neighborhood of this point. Let us indicate the plan of the proof of this theorem.

As is well-known, the *Dirichlet kernel*

$$D_n(x) = \sum_{|k|\le n} e^{ikx}$$

in the one-dimensional case has the form

$$D_n(x) = \frac{\sin(n+\frac{1}{2})x}{2\sin\frac{x}{2}}.$$

Let us fix a number δ, $0 < \delta < \pi$, and set $\Omega_\delta = \{x \in \mathbb{T}^1, |x| > \delta\}$. It follows then from the summability of f that the function $f(x+y)\cdot(\sin\frac{y}{2})^{-1}$ is summable in the domain Ω_δ. Therefore we have

$$\int_{\Omega_\delta} f(x+y)D_n(y)dy \to 0, \quad n\to\infty,$$

uniformly in $x \in [-\pi, \pi]$. Therefore the partial sum of the Fourier series of f is of the form

$$S_n(x,f) = \frac{1}{\pi}\int_{-\delta}^{\delta} f(x+y)D_n(y)dy + o(1),$$

from which the localization theorem follows.

Another formulation of the *localization principle* is: if two functions coincide in the neighborhood of a point, then the partial sums of their Fourier series have the same behavior there. Taking the difference of these functions we obtain an even more compact formulation: if f vanishes in a neighborhood of the point then $S_n(x,f) \to 0$ for $n\to\infty$.

In the multidimensional case the localization principle does not hold in the class $L_1(\mathbb{T}^n)$, neither for rectangular nor for spherical partial sums, which is connected with the unboundedness of the Dirichlet kernel of a neighborhood of the origin. In order to maintain localization one has to restrict the class of functions under consideration, by raising their smoothness. Then the classes are different depending on how the partial sums are formed.

1.2. Localization of Rectangular Partial Sums. It is easy to establish the non-existence of localization, say, for double Fourier series in the class of continuous functions $C(\mathbb{T}^2)$. It suffices to take (Tonelli (1928)) the function

$$f(x_1,x_2) = f_1(x_1)f_2(x_2),$$

where $f_1(x_1)$ vanishes for $|x_1| < \delta$ and $f_1 \in C(\mathbb{T})$ is such that the sum of partial sums $S_{m_1}(x_1, f_1)$ of its Fourier series at the point $x_1 = 0$ is different from zero for infinitely many indices m_1 (of course, then $S_{m_1}(0, f_1) \to 0$, $m_1 \to \infty$, by the classical Riemann localization principle). The second function $f_2 \in C(\mathbb{T})$ is chosen such that for $x_2 = 0$ the sequence of its partial sums $S_{m_2}(0, f_2)$ is unbounded (the existence of such a function was first proved by Du Bois-Reymond, cf. Bari (1961)). Then the function $f(x_1, x_2) \in C(\mathbb{T}^2)$ thereby defined vanishes in a δ-neighborhood of the origin $(0, 0)$ (and even in the strip $\{x \in \mathbb{T}^2 : |x_1| < \delta\}$), but the sequence of its rectangular partial sums

$$S_{m_1 m_2}(0, 0, f) = S_{m_1}(0, f_1) \cdot S_{m_2}(0, f_2)$$

at the point $(0, 0)$ is unbounded, which can be seen by choosing for each index m_1 such that $S_{m_1}(0, f_1) \neq 0$ the corresponding index m_2 sufficiently large.

This example extends in an obvious manner to an arbitrary number of dimensions $N > 2$. In general, we remark that every example of a function of $N - 1$ variables with a divergent Fourier series yields at the same time an example of a function in N variables with an analogous property.

The first attempt to find an acceptable analogue of the localization principle for rectangular sums was made by Tonelli (1928). It was based on the following observation: the rectangular partial sum $S_m(0, f)$ is an integral operator

$$S_m(x, f) = \frac{1}{\pi^n} \int_{\mathbb{T}^N} \tilde{D}_m(x - y) f(y) dy, \tag{1.1}$$

where

$$\tilde{D}_m(x) = \prod_{j=1}^{N} D_{m_j}(x_j), \tag{1.2}$$

$D_{m_j}(x_j)$ being the "usual" Dirichlet kernel. The singularity of $\tilde{D}_m(x)$ extends over the subspaces $x_j = 0$, the kernel $\tilde{D}$ being unbounded in a δ-neighborhood of these subspaces. Therefore, if $f(x) = 0$ in such δ-neighborhoods, then $S_m(x, f) \to 0$ as $\min m_j \to \infty$. In particular, Tonelli's theorem guarantees that the rectangular partial sums of the double sum at the point $a = (a_1, a_2)$ of a function $f \in L_1(\mathbb{T}^2)$ vanishing in a "cross-shaped" δ-neighborhood

$$U = \{x \in \mathbb{T}^2 : |x_1 - a_1| < \delta \quad \text{or} \quad |x_2 - a_2| < \delta\}$$

converge to zero.

Of course, this is not the principle of localization in its usual sense, as for the convergence of the partial sums of the Fourier series of the function f one has to impose a condition on f at points which are far away from the point under consideration. A natural localization principle requires that one passes from the class $L_1(\mathbb{T}^N)$ to smoother functions.

It has turned out that the Nikol'skiĭ classes $H_p^l(\mathbb{T}^N)$ (which are slightly larger than the Sobolev and Liouville classes) are most suitable for this purpose. In these classes V. A. Il'in (1970) has found the exact conditions for the localization of rectangular partial sums:

$$pl > N - 1, \quad l > 0, p \geq 1. \tag{1.3}$$

Namely, if (1.3) is fulfilled, then for any function $f \in H_p^l(\mathbb{T}^N)$, vanishing in some domain $\Omega \subset \mathbb{T}^N$, the rectangular partial sums $S_m(x, f)$ of the multiple Fourier series of f converge uniformly to zero on each compact set $K \subset \Omega$. If $pl \leq N - 1$ then there exists a function $f \in H_p^l(\mathbb{T}^N)$, vanishing in a neighborhood of the origin, such that

$$\overline{\lim_{\min m_j \to \infty}} |S_m(0, f)| > 0.$$

The proof in the N-dimensional case reduces in fact, thanks to a clever device, to the proof of the convergence on the $(N-1)$-dimensional torus. To this end, each function vanishing, say, at the origin is written in the form

$$f(x) = \sum_{j=1}^{N} f_j(x),$$

where $f_j(x) = 0$ in the strip $\{x \in \mathbb{T}^N : |x_j| < \delta\}$. If we replace f by f_j in (1.1), then the singularity of the integrand in the x_j-variable disappears. In fact, there occurs a smoothing in the x_j direction and the problem is reduced to computing the partial sums of an $(N-1)$-dimensional Fourier series of a function in $H_p^l(\mathbb{T}^{N-1})$, $pl > N-1$ (which however not necessarily must vanish near the origin). The last inequality (more about this in the next section) is a necessary condition for uniform convergence in $\mathbb{T}^{N-1}$, which then implies the statement about localization. Let us however point out that the rigorous proof of each of these steps involves considerable difficulties and technical complications, which have to be overcome.

The connection between localization on $\mathbb{T}^N$ and convergence on $\mathbb{T}^{N-1}$ of rectangular partial sums becomes especially apparent in the construction of so-called counterexamples. For example, let $g \in H_p^l(\mathbb{T}^{N-1})$ be a function such that its rectangular partial sums are unbounded at the point $x = 0$, where we do not assume that g vanishes in a neighborhood of the origin. Take $h \in C^\infty(\mathbb{T}^1)$ such that $h(t) = 0$ for $|t| < \delta$ and $S_n(0, h) \neq 0$ for infinitely many indices n. Then putting

$$f(x) = g(x_1, x_2, \ldots, x_{N-1}) \cdot h(x_N),$$

we get a function $f \in H_p^l(\mathbb{T}^N)$ (with the same p and l) for which localization fails.

The above argument illustrates anew the fact that each example of divergence in $\mathbb{T}^{N-1}$ gives an example of non-localization in $\mathbb{T}^N$. Let us point out that in the construction of counterexamples it is essential that we use the method of rectangular summation. In Goffman and Liu (1972) there is constructed a function in $W_p^1(\mathbb{T}^N)$, $p < N - 1$, for which the localization principle fails for square summation of the Fourier series.

An interesting generalization of the localization principle is given in Bloshanskiĭ (1975), where the following is proved.

If $f \in L_p(\mathbb{T}^2)$, $p > 1$, vanishes on some domain $\Omega \subset \mathbb{T}^2$ then the rectangular sum of f converges to 0 almost everywhere on Ω.

It is curious that this statement fails for $p = 1$, and does not extend to $L_p(\mathbb{T}^N)$ with $N > 2$. Since the generalized localization principle is closer to the problem of almost everywhere convergence, it will be discussed in greater detail in Chapter 3.

1.3. Localization of Spherical Partial Integrals. As a rule, spherical partial sums do not behave better than rectangular ones and, in particular, for them also the localization principle fails in $L_1(\mathbb{T}^N)$. Preservation of this class would require passing to classes of smoother functions, exactly as this was done for rectangular partial sums. However, it is much more effective to approach the regularization of spherical sums by way of Riesz means.

Before studying the partial sums of a Fourier series, let us consider the Riesz means of a Fourier integral:

$$R_\mu^s f(x) = (2\pi)^{-\frac{N}{2}} \int_{|\xi|<\mu} e^{ix\xi} \left(1 - \frac{|\xi|^2}{\mu^2}\right)^s \hat{f}(\xi)\, d\xi. \tag{1.4}$$

The operator R_μ^s is an integral operator with kernel

$$\theta^s(x-y,\mu) = (2\pi)^{-N} \int_{|\xi|<\mu} \left(1 - \frac{|\xi|^2}{\mu^2}\right)^s e^{i(x-y)\xi}\, d\xi. \tag{1.5}$$

For $s = 0$ this kernel is called the *spectral function of the Laplace operator* for the entire space $\mathbb{R}^N$. The Riesz means (1.5) of the spectral function can be computed explicitly by introducing polar coordinates, and take the form (cf. (47)):

$$\theta^s(x,\mu) = (2\pi)^{-N} 2^s \Gamma(s+1) \mu^{\frac{N}{2}-s} \frac{J_{\frac{N}{2}+s}(\mu|x|)}{|x|^{\frac{N}{2}+s}}. \tag{1.6}$$

In the introduction we remarked that the exponent $s = \frac{N-1}{2}$ was termed by Bochner (1936) the critical exponent. This appellation is, in particular, justified by the fact that the localization principle holds for the means (1.4) for $s \geq \frac{N-1}{2}$ but not for $s < \frac{N-1}{2}$. In order to prove this statement we shall turn to the identity (1.6). Using the estimate $|J_\nu(t)| \leq \mathrm{const} \cdot 1/\sqrt{t}$ for Bessel functions, we obtain

$$|\theta^s(x,\mu)| \leq \mathrm{const} \cdot |x|^{-\frac{N+1}{2}-s} \mu^{\frac{N-1}{2}-s}. \tag{1.7}$$

Assume that the function $f \in L_1(\mathbb{R}^N)$ vanishes in some domain $\Omega \subset \mathbb{R}^N$. For an arbitrary compact set $K \subset \Omega$, let $\delta = \mathrm{dist}(K, \partial\Omega)$. Then in the integral

$$R_\mu^s f(x) = \int_{\mathbb{R}^N \setminus \Omega} \theta^s(x-y,\mu) f(y) dy$$

for $x \in K$ the variable y will not appear if it is closer than δ to x, that is, we have $|x-y| > \delta$. In this case we obtain from the above formula and from (1.6)

$$|R_\mu^s f(x)| \leq \text{const} \cdot \mu^{\frac{N-1}{2}-s} \|f\|_{L_1}, \quad x \in K. \tag{1.8}$$

If $s > \frac{N-1}{2}$, it follows immediately from (1.8) that $R_\mu^s f \to 0$ for $\mu \to \infty$ uniformly on K, that is, the localization principle holds true. The proof is more complicated in the critical case $s = \frac{N-1}{2}$, because then the estimate (1.8) takes the form

$$|R_\mu^s f(x)| \leq \text{const} \cdot \|f\|_{L_1}, \quad x \in K. \tag{1.9}$$

The family of operators $R_\mu^s : L_1(\mathbb{R}^N) \to C(K)$ is uniformly bounded in the Banach space B of functions $f \in L_1(\mathbb{R}^N)$ which vanish on Ω, while it is convergent on the dense subset $C_0^\infty(\mathbb{R}^N)$. Therefore $R_\mu^s f \to 0$ uniformly on any compact set $K \subset \Omega$ for any $f \in B$, that is, the localization principle holds true also for $s = \frac{N-1}{2}$.

If however $s < \frac{N-1}{2}$ the localization principle breaks down. In fact, assume that for each function $f \in L_1(\mathbb{R}^N)$, which vanishes in some δ-neighborhood of the point $x = 0$, we have

$$R_\mu^s f(0) \to 0, \quad \mu \to \infty, s < \frac{N-1}{2}.$$

Then in view of the Banach-Steinhaus theorem for such functions we must have

$$|R_\mu^s f(0)| \leq \text{const} \cdot \|f\|_{L_1(\mathbb{R}^n)}, \quad x \in K. \tag{1.10}$$

Set $f(x) = \theta^s(x,\mu)$ if $\delta < |x| < 2\delta$ and $f(x) = 0$ otherwise. Then we obtain from (1.10)

$$\int_{\delta<|y|<2\delta} |\theta^s(y,\mu)|^2 dy \leq \text{const} \cdot \int_{\delta<|y|<2\delta} |\theta^s(y,\mu)| dy.$$

However, it follows from (1.6) and the asymptotics of Bessel functions that the left hand side of the last integral grows like μ^{N-1-2s}, while the right hand side behaves like $\mu^{\frac{N-1}{2}-s}$, which gives a contradiction to this inequality and, thus, also to (1.10), so that the localization principle cannot be true.

1.4. Equiconvergence of the Fourier Series and the Fourier Integral. Turning to the study of Riesz means of order s of spherical partial sums of multiple Fourier series,

$$\sigma_\mu^s(x,f) = \sum_{|n|<\mu} \left(1 - \frac{|n|^2}{\mu^2}\right)^s f_n e^{inx}, \tag{1.11}$$

let us fix our attention to the fact that the operator $f \mapsto \sigma_\mu^s(x,f)$ is an integral operator

$$\sigma_\mu^s(x,f) = \int_{\mathbb{T}^N} D_\mu^s(x-y)f(y)dy,$$

with kernel $D_\mu^s(x-y)$ (the Dirichlet kernel) given by

$$D_\mu^s(x-y) = (2\pi)^{-N} \sum_{|n|<\mu} \left(1 - \frac{|n|^2}{\mu^2}\right)^s e^{in(x-y)}. \tag{1.12}$$

If one compares the identities (1.12) and (1.5), one sees that the Fourier coefficient of the function $D_\mu^s(x)$ equals the value of the Fourier transform of $\theta^s(x,\mu)$ at the point $\xi = n$. Therefore, for $s > \frac{N-1}{2}$ we can apply Poisson's summation formula, yielding

$$D_\mu^s(x) = \sum_{n \in \mathbb{Z}^N} \theta^s(x + 2\pi n, \mu). \tag{1.13}$$

Equality (1.13) constitutes one of the main tools with the aid of which one studies spherical partial sums of multiple Fourier series. Let us split off the term with $n = 0$:

$$D_\mu^s(x) = \theta^s(x,\mu) + \rho^s(x,\mu), \tag{1.14}$$

where

$$\rho^s(x,\mu) = \sum_{n \neq 0} \theta^s(x + 2\pi n, \mu). \tag{1.15}$$

Estimating each of the terms in the sum (1.15) with the aid of formula (1.7), we obtain

$$|\rho^s(x,\mu)| \leq \text{const} \cdot \mu^{\frac{N-1}{2}-s} \sum_{n \neq 0} |x + 2\pi n|^{-\frac{N+1}{2}-s}.$$

If $s > \frac{N-1}{2}$ the series to the right is uniformly convergent for $x \in \{x \in \mathbb{R}^N : |x_j| \leq 2\pi - \delta\}$. Therefore we can rewrite the equality (1.14) as follows:

$$D_\mu(s) = \theta^s(x,\mu) + O(1) \cdot \mu^{\frac{N-1}{2}-s}, \quad s > \tfrac{N-1}{2}. \tag{1.16}$$

Assume that $f \in L_1(\mathbb{T}^N)$ vanishes near the boundary of $\mathbb{T}^N$. Its continuation by zero off $\mathbb{T}^N$ will likewise be denoted by f. Then we obtain from (1.16)

$$\sigma_\mu^s f(x) = R_\mu^s f(x) + O(1) \cdot \mu^{\frac{N-1}{2}-s} \|f\|_{L_1}, \quad s > \tfrac{N-1}{2}. \tag{1.17}$$

This equality shows that for $s > \frac{N-1}{2}$ we have uniform equiconvergence of the Riesz means of the Fourier series and the Fourier integral expansion.

This implies, in particular, the validity of the localization principle in the class $L_1(\mathbb{T}^N)$ of the means (1.11) with $s > \frac{N-1}{2}$.

Let us remark that relation (1.16) has been obtained only under the hypothesis $s > \frac{N-1}{2}$. Therefore it is not possibly to conclude anything from it for Riesz means of critical index. In particular, it does not allow us to prove the localization of the means (1.11) for $s = \frac{N-1}{2}$. However, the reason for this is not the weak equiconvergence of Poisson's formula but rather the essentially different behavior of spherical means at the critical index for Fourier series and Fourier integral expansion.

We have seen above that the localization of the means (1.4) holds for $s = \frac{N-1}{2}$. It is therefore rather surprising that for the means (1.11) the localization fails for $s = \frac{N-1}{2}$ (Bochner (1936)). This means that in the critical case the second term to the right in (1.14) begins to dominate and the equiconvergence disappears. This example is yet another illustration of the general truth that it is seldom possible to establish equidivergence with the aid of the equiconvergence theorem.

1.5. Riesz Means Below the Critical Index. If $s \leq \frac{N-1}{2}$, then for the localization of the Riesz means (1.11) it is necessary to require appropriate smoothness of the function at hand. The most complete results hold for Hilbert spaces of smooth functions (i. e. functions admitting square summable derivatives). We give the corresponding formulation for the Liouville classes $L_2^l(\mathbb{T}^N)$.

If

$$l + s \geq \frac{N-1}{2}, \quad l \geq 0,\ s \geq 0, \tag{1.18}$$

then the localization principle for the Riesz means (1.11) of order s holds in the classes $L_2^l(\mathbb{T}^N)$.

If $s = 0$ (i.e. for the partial sums themselves) this result was first established by V. A. Il'in (1957). In fact, the localization principle holds true in the slightly larger Nikol'skiĭ class H_2^l (cf. Il'in and Alimov (1971)).

As $L_p^l(\mathbb{T}^N) \subset L_q^l(\mathbb{T}^N)$ for $p > q$, the condition (1.18) entails localization also in the classes $L_p^l(\mathbb{T}^N)$ for $p \geq 2$. The extension of these results to the case $1 \leq p < 2$ is an extremely difficult problem. A partial solution of this problem can be found in the paper Bastis (1983), where it is shown that in the assumption (1.18) localization is in force also for $p > 2 - \varepsilon(N)$, where $\varepsilon(2) = \frac{7}{17}$, $\varepsilon(3) = \frac{2}{3}$, $\varepsilon(N) = \frac{2}{N+1}$ for $N \geq 4$. One can show (as is done in Bastis (1983)) that the inequality $l + s \geq \frac{N-1}{2}$ guarantees the localization in $L_p^l(\mathbb{T}^N)$ also for $1 < p \leq 2 - \varepsilon(N)$, but this succeeds not for all $s \geq 0$ but only for s close to $\frac{N-1}{2}$.

There arises the question how exact the condition (1.18) is and, in particular, whether it is possible to weaken it on the expense of increasing p. The following *theorem of Il'in* (cf. Il'in (1968)) *on the non-existence of localization* gives an answer to this question.

Theorem (V. A. Il'in). *Assume that $s \geq 0$, $l \geq 0$ and $l + s < \frac{N-1}{2}$. Then for each point $x^0 \in \mathbb{T}^N$ there exists a finite*[7] *function $f \in C^l(\mathbb{T}^N)$ satisfying the following conditions:*

1) $f(x) = 0$ in a neighborhood of the point x^0;

2) $\overline{\lim}_{\mu\to\infty} |\sigma^s_\mu(x^0, f)| = +\infty$.

This theorem means that for $l + s < \frac{N-1}{2}$ the localization in $L^l_p(\mathbb{T}^N)$ cannot be saved even if we pass to $p = \infty$, i. e. the sufficient condition (1.18) is in fact independent of p (at any rate, for $p \geq 2$).

The proof of Il'in's theorem is based on an estimate of the L_1-norm of the Riesz means of the spectral function (1.12):

$$\int_E |D^s_\mu(x)|dx \geq \alpha \cdot \mu^{\frac{N-1}{2}-s}, \tag{1.19}$$

where E is an arbitrary subset of $\mathbb{T}^N$ of sufficiently large measure, lying off a certain δ-neighborhood of the origin. Let us indicate how the divergence of the multiple Fourier series expansion of the function $f \in L^l_p(\mathbb{T}^N)$ follows from this estimate in the case $s = 0$ (if $s > 0$ the argument differs in some technical details). Let us denote by $L_p(E)$ the set of functions in $L_p(\mathbb{T}^N)$ which vanish on E. By definition (cf. (58)) for each $g \in L_p(E)$ the function

$$f(x) = \sum_{n\in\mathbf{Z}^N} (1 + |n|^2)^{-\frac{l}{2}} g_n e^{inx} \tag{1.20}$$

belongs to $L^l_p(\mathbb{T}^N)$. The spherical partial sums of the Fourier series of f and g are connected by the formula

$$E_\mu f(x) = \int_0^\mu (1 + t^2)^{-\frac{l}{2}} dE_t g(x).$$

Here $f \mapsto E_\mu f$ is an integral operator whose kernel $K_\mu(x-y)$ is connected with the Dirichlet kernel $D_\mu(x-y)$ by the formula

$$K_\mu(x) = \int_0^\mu (1 + t^2)^{-\frac{l}{2}} dD_t(x). \tag{1.21}$$

This means that

$$E_\mu f(x) = \int_E K_\mu(x-y)g(y)dy. \tag{1.22}$$

The convergence of $E_\mu f(0)$ for all $f \in L^l_p(\mathbb{T}^N)$ of the above type would imply, in view of the Banach-Steinhaus theorem, the boundedness of the

[7] *Translator's Note.* That is, a function with compact support. This terminology is, regretfully, rarely used in the Western literature.

kernel $K_\mu(x)$ in the $L_q(E)$ metric, where $\frac{1}{q} = 1 - \frac{1}{p}$. But then it follows from (1.21) that the norm of the kernel $D_\mu(x)$ in $L_q(E)$ is majorized by μ^l, which for $l < \frac{N-1}{2}$ and $s = 0$ contradicts Il'in's estimate (1.19).

Consequently, there exists a function $f \in L_p^l(\mathbb{T}^N)$ of the form (1.20), where $g \in L_p(E)$, with a divergent Fourier series at the origin for spherical summation. With the aid of this function f it is also not hard to construct a function satisfying all the conditions of Il'in's theorem.[8]

In the last step one uses the property of pseudo-localization of the operator $g \mapsto f$ defined in (1.20). This property means that the operator at hand does not increase the singular support, i.e. if $g \in C^\infty(\Omega)$, $\Omega \subset \mathbb{T}^N$ then $f \in C^\infty(\Omega)$ (cf. Hörmander (1983–85), Taylor (1981), Chapter 2, Sect. 2).

1.6. Localization Under Summation over Domains Which Are Level Sets of an Elliptic Polynomial. Let $A(\xi)$ be an arbitrary homogeneous elliptic polynomial (cf. (15)) and set

$$E_\lambda^s f(x) = \sum_{A(n)<\lambda} \left(1 - \frac{A(n)}{\lambda}\right)^s f_n e^{inx}. \tag{1.23}$$

For orientation let us remark that the behavior of the means (1.23) depends in an essential way on the geometry of the surface

$$\partial Q_A = \{\xi \in \mathbb{R}^N : A(\xi) = 1\}, \tag{1.24}$$

more exactly, on the number r of non-vanishing principal curvatures of this surface. In view of this criterion we may make a classification of homogeneous elliptic operators, referring to one class A_r all operators such that each point of the surface (1.24) at least r of the $N-1$ principal curvatures are different from zero. It is clear that

$$A_{N-1} \subset \cdots \subset A_1 \subset A_0,$$

where A_0 is the class of all homogenous elliptic operators. A typical representative of the class A_{N-1} is the Laplace operator. As an example of an elliptic operator in A_r, $r < N-1$, we may take

$$M_r(D) = \left(\sum_{j=1}^{r+1} D_j^2\right)^{m+1} + \left(\sum_{j=r+2}^{N} D_j^2\right)^m \left(\sum_{j=1}^{N} D_j^2\right);$$

note that $M_r \notin A_{r+1}$.

[8] In view of the imbedding theorem (59) we have $L_p^l(\mathbb{T}^N) \to C^{l-\varepsilon}(\mathbb{T}^N)$ for all $\varepsilon > 0$, provided $p = p(\varepsilon)$ is sufficiently small.

In the class A_0 the exact sufficient condition for the localization of Riesz means of order s of multiple Fourier series and integrals in the class L_p^l has the form (Alimov (1974)):

$$l + s \geq \tfrac{N-1}{p}, \quad l \geq 0, \quad s \geq 0, \quad 1 \leq p \leq 2 \tag{1.25}$$

(if $l = 0$ this result is due to Hörmander (1969)). If $p > 2$ then an unimprovable condition is (1.18), whose union with (1.25) can be written

$$l + s \geq \max\{\tfrac{N-1}{p}, \tfrac{N-1}{2}\}, \quad 1 \leq p \leq 2.$$

For operators in the narrower classes A_r condition (1.25) can be weakened to (provided the set Q_A is convex):

$$l + s \geq \tfrac{N-1}{p} - r(\tfrac{1}{p} - \tfrac{1}{2}), \quad 1 \leq p \leq 2. \tag{1.26}$$

The conditions (1.26) are sufficient for the localization of the expansion in multiple Fourier integrals (Ashurov (1983a)). Concerning the expansion in multiple Fourier series, one has so far only been able to prove the sufficiency of these conditions for $l = 0$ (Ashurov (1985)).

In which sense is the condition (1.26) exact? For each fixed operator in A_r (with $r < N - 1$) one can weaken them, but within the entire class A_r they are exact (Ashurov (1983a)).

Namely, assume that

$$l + s < \tfrac{N-1}{p} - r(\tfrac{1}{p} - \tfrac{1}{2}), \quad 1 \leq p \leq 2.$$

Then there exist an elliptic operator $L \in A_r$ such that the localization principle does not hold true for the Riesz means of the expansion corresponding to it. This result, as well as the sufficiency of (1.26), has been proved in the general form for Fourier integrals and, if $l = 0$, for Fourier series. For $r = 0$ we may take for L the operator (34) in Pulatov (1981).

§2. Uniform Convergence

2.1. The One-Dimensional Case. The problem of convergence is closely connected with the localization principle for the expansions. If the function under examination belongs to a class for which the localization principle holds true, then the convergence or divergence at a given point is determined by the smoothness of f in just a neighborhood of that point.

In the one-dimensional case the localization principle holds in $L_1(\mathbb{T}^1)$, so that for any $f \in L_1(\mathbb{T}^1)$ the convergence of its Fourier series at the point x^0 depends only on the smoothness of f near x^0. The required smoothness condition comes in integral form and is known as the *Dini condition*:

$$\int_0^\delta |f(x^0 + t) - 2f(x^0) + f(x^0 - t)| \frac{dt}{t} < \infty. \tag{1.27}$$

For the uniform convergence of the Fourier series in an arbitrary interval I it suffices to require that the integral (1.27) converges uniformly for x^0 in that interval. On the other hand, the well-known *Dini-Lipschitz test* for uniform convergence on $\mathbb{T}^1$ does not assume that (1.27) is fulfilled. Let us recall this classical test, for which purpose we introduce the notion of *modulus of continuity*:

$$\omega(\delta, f) = \sup_{|x-y|<\delta} |f(x) - f(y)|. \tag{1.28}$$

In the right hand side of (1.28) the least upper bound is taken over all points $x, y \in \mathbb{T}^1$ such that $|x - y| < \delta$. If

$$\omega(\delta, f) = o(1)\left(\log \frac{1}{\delta}\right)^{-1}, \quad \delta \to 0, \tag{1.29}$$

then the Fourier series of f converges to f uniformly on $\mathbb{T}^1$.

As is well-known (cf. Bari (1961)), the continuity only of f is not sufficient for the uniform convergence of the Fourier series of f and, more generally, not even for the convergence at a fixed point.

Indeed, if we consider $S_n(x, f)$ as an operator from $C(\mathbb{T}^1)$ into itself, then its norm equals

$$\sup_{\|f\|_C \leq 1} \sup_{x \in \mathbb{T}^1} \left| \int_{\mathbb{T}^1} f(x+t) D_n(t) dt \right| = \int_{\mathbb{T}^1} |D_n(t)| dt,$$

i.e. the *Lebesgue constant*, which grows as $\log n$ as $n \to \infty$.

Moreover, the divergence set may have the cardinality of the continuum (although it always is a set of zero measure, as follows from Carleson's theorem (1966)). Therefore it follows that auxiliary conditions of the Dini-Lipschitz type cannot be dispensed with.

The Dini-Lipschitz condition is exact in the sense that $o(1)$ cannot be replaced by $O(1)$. Moreover, the condition $f(x^0 + t) - f(x^0) = o(1)(\log \frac{1}{t})^{-1}$ does not gurantee the convergence of the Fourier series of f at the point x^0, that is, condition (1.29) is only a test for uniform convergence.

2.2. Rectangular Sums. A generalization of the theorem of Dini-Lipschitz to the case of rectangular partial sums for $N > 1$ was given by L. V. Zhizhiashvili (1971), (1973). He show that if

$$\omega(\delta, f) = o(1)\left(\log \frac{1}{\delta}\right)^{-N}, \quad \delta \to 0, \tag{1.30}$$

then the Fourier series converges uniformly to f on$\mathbb{T}^N$, provided we use rectangular summation. The modulus of continuity in (1.30) is again defined by (1.28) but now with $x \in \mathbb{T}^N$, $y \in \mathbb{T}^N$.

Condition (1.30) is sharp, and also in the case of quadratic summability it cannot be improved: there exists a function f such that

$$\omega(\delta, f) = O(1)\left(\log\frac{1}{t}\right)^{-N}, \quad \delta \to 0,$$

but whose Fourier series diverges at the point $x = 0$ under quadratic summation.

The proof of the necessity of condition (1.30) is based on the fact that the Dirichlet kernel $D_m(x)$ for rectangular partial sums is a product of one-dimensional Dirichlet kernels (cf. (1.2)).

As

$$\pi^{-N}\int_{\mathbf{T}^N} D_m(x)dx = 1,$$

we have

$$S_m(x, f) - f(x) = \pi^{-N}\int_{\mathbf{T}^N}[f(x+y) - f(y)]D_m(y)dy. \tag{1.31}$$

Next, let us note that one has the following estimate for the rate of convergence in the one-dimensional case, which can be obtained upon a careful analysis of the proof of the Dini-Lipschitz theorem: if

$$\omega(\delta, f) = o(1)\left(\log\frac{1}{\delta}\right)^{-\lambda}, \quad \lambda \geq 1,\ \delta \to 0$$

then uniformly on $\mathbb{T}^1$

$$S_n(x, f) - f(x) = o(1)(\log n)^{-\lambda+1}, \quad n \to \infty.$$

Furthermore, one has to represent the difference $f(x+y)-f(x)$ in (1.31) as a sum of increments with respect to each coordinate x_k. Integrating the term with index k with respect to x_j, we note that for $j < k$ we integrate effectively only the Dirichlet kernel $D_{m_j}(y_j)$ (the integral of which equals π) and that for $j = k$ we can use the aforementioned sharpening of the Dini-Lipschitz theorem and for $j > k$ the estimate

$$\int_{-\pi}^{\pi} |D_{m_j}(y_j)|dy_j \leq c\log m_j.$$

Hence we obtain

$$S_m(x, f) - f(x) = o(1)\cdot\sum_{k=1}^{N}(\log m_k)^{-N+1}\prod_{j=k+1}^{N}\log m_j = o(1).$$

Here we may assume, without loss of generality, that $m_1 \leq m_2 \leq \cdots \leq m_N$.

2.3. Uniform Convergence of Spherical Sums. In contrast to rectangular partial sums, for the uniform convergence of spherical partial sums one has to require quite a lot of smoothness (compared to the one-dimensional case). The necessary condition for uniform convergence with respect to rectangles (1.30) hardly differs from the condition for continuity and is further not detected by the power smoothness scale, i.e. the usual Hölder classes are too crude for the description of the situation at hand. This is connected with the slow (logarithmic) growth of the $L_1(\mathbb{T}^N)$-norm of the Dirichlet kernel (or, more exactly, the Lebesgue constant) for the rectangular sums. In fact,

$$L_m = \int_{\mathbb{T}^N} |D_m(x)|dx = \prod_{k=1}^{N} \int_{\mathbb{T}^1} |D_{m_k}(x_k)|dx_k = \prod_{k=1}^{N} \log m_k \cdot (1 + o(1)).$$

A completely different situation is encountered in spherical summation. In the $L_1(\mathbb{T}^N)$-norm the corresponding Dirichlet kernel grows like a power, so that for convergence one requires high smoothness (in classical or in the generalized sense). If the function under examination does not display high smoothness then we must resort to averages of the partial sums, usually the Riesz averages. For the expansion in a Fourier integral this follows at once from formula (1.6):

$$L_\lambda \equiv \int_{\mathbf{R}^N} |\theta^s(x, \lambda)|dx = \text{const} \cdot \log \lambda \, (1 + o(1)), \quad s = \tfrac{N-1}{2}, \, \lambda \to \infty.$$

And once again, as in the localization problem, one is faced with the critical exponent $s = \frac{N-1}{2}$.

Bochner (1936) showed that if $s > \frac{N-1}{2}$ then the Riesz mean of order s (cf. (19)) converge for any continuous function to the same function and this uniformly on $\mathbb{T}^N$. But if $s = \frac{N-1}{2}$ this is not true anymore. In this sense the Riesz means at the critical exponent behave as the partial sums of the one-dimensional Fourier series: for their uniform convergence continuity alone does not suffice and it is necessary to impose "logarithmic" smoothness. Namely, if the modulus of continuity of $f \in C(\mathbb{T}^N)$ satisfies the Dini-Lipschitz condition

$$\omega(\delta, f) = o(1) \log \tfrac{1}{\delta}, \quad \delta \to 0,$$

then the Riesz means of f of critical exponent converge to f uniformly on $\mathbb{T}^N$ (cf. Golubov (1975)).

As in the one-dimensional case the *convergence test of Dini* holds true. In order to formulate the corresponding result let us make the following definition: the function $f \in C(\mathbb{T}^N)$ satisfies the *Dini condition* at the point $x \in \mathbb{T}^N$, if for some $\delta > 0$ the integral

$$\int_0^\delta |f_r(x) - f(x)| \frac{dr}{r}$$

is convergent, where $f_r(x)$ denotes the average value of f over a sphere of radius r centered at the point x.

If f satisfies a Dini condition at the point x, then its Riesz means of critical order at this point converge to $f(x)$ (Bochner (1936)). If the Dini condition is fulfilled uniformly on $\mathbb{T}^N$, then the convergence is uniform on $\mathbb{T}^N$.

We have already remarked that the sufficient conditions for the uniform convegence of the spherical partial sums, as well as of their Riesz means of order lower than the critical, require increased smoothness of the function under view. We formulate these conditions in the Liouville classes $L_p^l(\mathbb{T}^N)$, $p \geq 1$, $l > 0$:

$$l + s \geq \tfrac{N-1}{2}, \quad l \cdot p > N,\ s \geq 0. \tag{1.32}$$

If (1.32) is fulfilled then the Fourier series of any function $f \in L_p^l(\mathbb{T}^N)$ is summable to it by Riesz means of order s, uniformly on $\mathbb{T}^N$ (cf. Il'in and Alimov (1971)). The first of the inequalities (1.32) is specially important. That it is not possible to weaken it in the Hölder class C^l follows from V. A. Il'in's theorem on the non-existence of localization (cf. Sect. 1.5).

The second inequality $lp > N$ is also essential, at least to make sure that in the hypothesis of the opposite inequality $lp \leq N$ there exists an unbounded function in L_p^l whose Fourier series trivially is divergent and cannot be uniformly summable on $\mathbb{T}^N$ (for $lp > N$ one has the imbedding $L_p^l(\mathbb{T}^N) \to C(\mathbb{T}^N)$). However, this is not the whole story: the inequality $lp > N$ guarantees that all functions in $L_p^l(\mathbb{T}^N)$ are Hölder continuous with exponent $\alpha = l - \frac{N}{p}$. That one cannot ignore the Hölder condition can be concluded from the following theorem.

Theorem (Alimov (1978)). *Assume that $s \geq 0$, $p \geq 1$ and let l be an integer such that*

$$l + s = \frac{N-1}{2}, \quad lp = N.$$

Then one can find for each point $x^0 \in \mathbb{T}^N$ a function $f \in W_p^l(\mathbb{T}^N) \cap C(\mathbb{T}^N)$ such that

$$\varlimsup_{\lambda \to \infty} |E_\lambda^s f(x^0)| = +\infty.$$

If $l + s$ exceeds the critical exponent no additional smoothness is required and the uniform convergence holds true for any $f \in W_p^l(\mathbb{T}^N) \cap C(\mathbb{T}^N)$, $lp = N$.

2.4. Summation over Domains Bounded by the Level Surfaces of an Elliptic Polynomial. In Sect. 1.16 we remarked that the conditions for localization in the class L_p^l of partial sums depend in an essential way on the properties of the principal symbol $A_0(\xi)$ of an elliptic operator A, more precisely on the degree of convexity of the set

$$Q_A = \{\xi \in \mathbb{R}^N : A_0(\xi) < 1\}.$$

Therefore it seems, at the first glance, quite unexpected that the conditions for uniform convergence in L_p^l, $l > 0$, do not depend on the symbol of the operator, and take for all elliptic operators exactly the same form as for the Laplace operator (cf. Alimov (1973)). As the level surface ∂Q_A for the symbol of Laplace's operator is a sphere, these conditions coincide with the conditions (1.32) for uniform convergence of spherical partial sums.

One can explain this difference between uniform convergence and localization by noting that the most significant of the conditions (1.32) is the "limit" case $l + s = \frac{N-1}{2}$, $lp = N + \varepsilon$, $\varepsilon > 0$. The remaining cases can be reduced to this case by applying the imbedding theorems, along with the following property of Riesz means: if the means of order s converge, then the means of order $s' > s$ are likewise convergent with the same limit. Now, the limit case appears only if $p > \frac{2N}{N-1} > 2$, and therefore it is possible to compare (1.32) with the conditions for localization (cf. (1.26))

$$l + s \geq \max\left\{\frac{N-1}{2}, \frac{N-1}{p} - r\left(\frac{1}{p} - \frac{1}{2}\right)\right\}$$

for $p \geq 2$, that is, for those p such that the conditions for localization do not depend on the symbol of the operator.

The sharpness of condition (1.32) for all elliptic operators follows from the fact that it is sharp for the Laplace operator.

Chapter 2
L_p-Theory

§1. Convergence of Fourier Series in L_p

1.1. Rectangular Convergence. By the classical theorem of M. Riesz the one-dimensional Fourier series of a function $f \in L_p(\mathbb{T}^1)$ converges to this function in the metric of $L_p(\mathbb{T}^1)$ provided $1 < p < \infty$. Otherwise put, the system $\{e^{inx}\}_{n=-\infty}^{\infty}$ constitutes a *basis* in $L_p(\mathbb{T}^1)$ for $1 < p < \infty$. In the extremal cases $p = 1$ and $p = \infty$ this is not true.

For rectangular summation the basis property in $L_p(\mathbb{T}^N)$, $1 < p < \infty$, remains in force for any $N > 1$, as was proved by Sokol-Sokolowski (1947). In fact, the following more general statement is true: if the operators $P_n : L_p(\mathbb{T}^1) \to L_p(\mathbb{T}^1)$ and $Q_n : L_p(\mathbb{T}^1) \to L_p(\mathbb{T}^1)$ are uniformly bounded, then the operators

$$R_{nm} f(x, y) = P_n Q_m f(x, y),$$

where P_n acts on the variable $x \in \mathbb{T}^1$ and Q_m on the variable $y \in \mathbb{T}^1$, are likewise uniformly bounded. As a system of functions is a basis if and only if the operators of partial sums are uniformly bounded, this implies the theorem of Sokol-Sokolowski.

In the study of bases it is often useful to make the following observation. Let $\{v_n\}_{n=1}^{\infty}$ be an orthonormal system in $L_2(\mathbb{T}^1)$ and denote by S_n the operator of partial sum:

$$S_n(f,x) = \sum_{k=1}^{n}(f, v_k)v_k(x).$$

Usually one takes $v_k \in L_\infty(\mathbb{T}^1)$. Then S_n extends in a natural way to $L_p(\mathbb{T}^1)$, $p \geq 1$. Let us assume that the operators S_n are uniformly bounded in L_p. As $S_n^* = S_n : L_q \to L_q$ and

$$\|S_n^*\|_{L_q \to L_q} = \|S_n\|_{L_p \to L_p},$$

they are bounded in L_q, $\frac{1}{p} + \frac{1}{q} = 1$.

From this statement it follows that if the system $\{v_n\}_{n=1}^{\infty}$ is a basis in $L_p(\mathbb{T}^1)$ then it is also a basis in $L_q(\mathbb{T}^1)$ and, therefore, in view of the Riesz interpolation theorem, a basis in $L_r(\mathbb{T}^1)$, where r is any number between p and q, $\frac{1}{p} + \frac{1}{q} = 1$.

We have already mentioned (cf. Chapter 1, Sect. 1.2) that there exists a continuous function whose Fourier series is divergent at some point (which is a consequence of the unboundedness of the Lebesgue constant). This means that the trigonometric system is not a basis in $C(\mathbb{T}^1)$, and not in $L_\infty(\mathbb{T}^1)$ and, by duality, not in $L_1(\mathbb{T}^1)$. As every function in one variable can be viewed as a function of N variables, the multiple series do not give rise to a basis in $L_\infty(\mathbb{T}^N)$ and $L_1(\mathbb{T}^N)$, that is, the strict inequality $1 < p < \infty$ is necessary in the Sokol-Sokolowski theorem.

1.2. Circular Convergence. The fact that the property of being a basis of the multiple trigonometric system depends on the way how the partial sums are formed, forces us to give a new definition of basis. Let $\Omega = \{\omega\}$ be one of the families of sets, considered in Sect. 3 of the introduction, defining a method of summation of multiple series. We say that the multiple trigonometric system is a *basis* for Ω in $L_p(\mathbb{T}^N)$ if for every function $f \in L_p(\mathbb{T}^N)$ its Fourier series converges to it with respect to Ω in the $L_p(\mathbb{T}^N)$ metric.

Let us consider from this point of view the question of basis in $L_2(\mathbb{T}^N)$. It follows immediately from Parseval's formula that every complete orthonormal system in a Hilbert space H is an *unconditional basis* in H (i. e. it remains a basis after an arbitrary permutation of its elements). Consequently, the system $\{e^{inx}\}_{n\in\mathbb{Z}^N}$ is an unconditional basis in $L_2(\mathbb{T}^N)$, and therefore the multiple series of any function $f \in L_2(\mathbb{T}^N)$ converges to f in the $L_2(\mathbb{T}^N)$-metric with respect to any method of summation: with respect to rectangles, balls, more generally, any system of sets exhausting $\mathbb{Z}^N$.

For $p \neq 2$ the state of affairs is completely different. Indeed, if $p \neq 2$ then the trigonometric system is not a basis for spherical summation. This follows from Fefferman's theorem on spherical multipliers (cf. the following Sect. 2). Namely, if the spherical partial sums

$$S_\mu(f,x) = \sum_{|n|\leq\mu} f_n e^{inx}$$

for every function $f \in L_p(\mathbb{T}^N)$ converge to f in the $L_p(\mathbb{T}^N)$-metric then, by the Banach-Steinhaus theorem, we must have

$$\|S_\mu f\|_{L_p(\mathbb{T}^N)} \leq M\|f\|_{L_p(\mathbb{T}^N)}. \tag{2.1}$$

From this it is not hard to deduce that the same inequality (with the same constant M) holds in any cube

$$\mathbb{T}_h^N = \{x \in \mathbb{R}^N : |x_j| \leq \frac{\pi}{h}\}.$$

Letting $h \to 0$ we obtain it for the spherical means for the expansion in a Fourier integral, which contradicts Fefferman's theorem (1970).

Let us present this reasoning in greater detail (cf. Mityagin and Nikishin (1973a)).

The orthonormal trigonometric system in $\mathbb{T}_h^N$ is $\{(2\pi)^{-N/2}h^{N/2}e^{ihnx}\}_{n\in\mathbb{Z}^N}$. Introduce the corresponding Dirichlet kernel

$$D_\mu^h(x) = (2\pi)^{-N}h^N \sum_{|n|\leq\mu} e^{ihnx}. \tag{2.2}$$

Let $f \in C_0^\infty(\mathbb{R}^N)$ and let $h > 0$ be a number so small that the support of f contains $\mathbb{T}_h^N$. Then f can be expanded in a multiple Fourier series in $\mathbb{T}_h^N$, and the spherical partial sum equals

$$S_\mu^h f(x) = \int_{\mathbb{T}_h^N} D_\mu^h(x-y)f(y)dy. \tag{2.3}$$

Put $f_h(x) = f(x/h)$. Then it follows from (2.2) and (2.3) that $S_\mu^h f(x) = S_\mu f_h(hx)$, which, in view of (2.1), gives

$$\|S_\mu^h f\|_{L_p(\mathbb{T}_h^N)} = h^{-N/p}\|S_\mu^h f_h\|_{L_p(\mathbb{T}_h^N)} \leq Mh^{-N/p}\|f_h\|_{L_p(\mathbb{T}^N)} = M\|f\|_{L_p(\mathbb{T}_h^N)}.$$

Consequently, for any $f \in C_0^\infty(\mathbb{R}^N)$ and for h sufficiently small

$$\|S_{\mu/h}^h f\|_{L_p(\mathbb{T}_h^N)} \leq M\|f\|_{L_p(\mathbb{R}^N)}. \tag{2.4}$$

Let now h tend to 0. It follows from (2.2) and (2.3) in an obvious manner that

$$S^h_{\mu/h}f(x) \to \sigma^s_\mu f(x) \equiv (2\pi)^{-N/2} \int_{|\xi|\leq\mu} \hat{f}(\xi)e^{ix\xi}d\xi.$$

Therefore, we obtain from (2.4) the inequality

$$\|\sigma^s_\mu f\|_{L_p(\mathbf{R}^N)} \leq M\|f\|_{L_p(\mathbf{R}^N)}, \tag{2.5}$$

which for $p \neq 2$ is not true, by Fefferman (1970).

1.3. Summation of Spherical Riesz Means. The argument in the previous section shows that for the approximation of functions in the L_p-metric by multiple series it is, for $p \neq 2$, necessary to consider means of the partial sums.

Bochner's result (1936) on uniform convergence (cf. Sect. 2 of Chapter 1) shows that the Riesz means of order $s > \frac{N-1}{2}$ for any continuous function converge to it in the $L_\infty(\mathbb{T}^N)$-metric. In this case it follows from the duality principle that for $s > \frac{N-1}{2}$, we have $E^s_\lambda f \to f$ $(\lambda \to \infty)$ in the $L_1(\mathbb{T}^N)$-metric for any $f \in L_1(\mathbb{T}^N)$, and therefore

$$\|E^s_\lambda f\|_{L_1(\mathbf{T}^N)} \leq M\|f\|_{L_1(\mathbf{T}^N)}, \quad s > \frac{N-1}{2}. \tag{2.6}$$

It follows from the orthogonality of the trigonometric system that for each $f \in L_2(\mathbb{T}^N)$ holds the estimate

$$\|E^s_\lambda f\|_{L_2(\mathbf{T}^N)} \leq M\|f\|_{L_2(\mathbf{T}^N)}, \quad s \geq 0. \tag{2.7}$$

It is intuitively clear that for intermediate p, i. e. $1 < p < 2$, an analogous estimate must hold with an s that decreases from $\frac{N-1}{2}$ to 0. That this is the case follows from Stein's interpolation theorem (Stein and Weiss (1971)). Ordinary interpolation (à la M. Riesz) allows one to interpolate the inequalities (2.6) and (2.7) for a fixed operator, i. e. it does not provide the possibility to change (in our case to lower) the order s when passing from L_1 to L_2. However, if the dependence of the operators under consideration on the parameter s is analytic (for this it is necessary to invoke Riesz means of complex order s) then one can carry out the interpolation in s.

Let us state *Stein's interpolation theorem* in a form suitable for our purposes.

We say that a function $\varphi(\tau)$, $\tau \in \mathbb{R}^1$, has admissible growth if there exist constants $a < \pi$ and $b > 0$ such that

$$|\varphi(z)| \leq \exp(b \exp a|\tau|). \tag{2.8}$$

Let A_z be a family of operators defined for *simple functions* (i. e. functions which are finite linear combinations of characteristic functions of measurable subsets of $\mathbb{T}^N$). We term the family A_z admissible if for any two simple functions f and g the function

$$\varphi(z) = \int_{\mathbb{T}^N} f(x) A_z g(x) dx$$

is analytic in the strip $0 \le \operatorname{Re} z \le 1$ and has admissible growth in $\operatorname{Im} z$, uniformly in $\operatorname{Re} z$ (this means that we have an estimate in $\operatorname{Im} z$ which is analogous to (2.8), with constants a and b independent of $\operatorname{Re} z$).

Theorem. *Let A_z be an admissible family of linear operators such that*

$$\|A_{i\tau} f\|_{L_{p_0}(\mathbb{T}^N)} \le M_0(\tau) \|f\|_{L_{p_0}(\mathbb{T}^N)}, \quad 1 \le p_0 \le \infty,$$

$$\|A_{1+i\tau} f\|_{L_{p_1}(\mathbb{T}^N)} \le M_1(\tau) \|f\|_{L_{p_1}(\mathbb{T}^N)}, \quad 1 \le p_1 \le \infty,$$

for all simple functions f and with $M_j(\tau)$ independent of τ and of admissible growth. Then there exists for each t, $0 \le t \le 1$, a constant M_t such that for every simple function f holds

$$\|A_t f\|_{L_{p_t}(\mathbb{T}^N)} \le M_t \|f\|_{L_{p_t}(\mathbb{T}^N)}, \quad \frac{1}{p_t} = \frac{1-t}{p_0} + \frac{t}{p_1}.$$

Some of the most useful objects (but by far not the only one) to which this theorem can be applied are the Riesz means, which analytically depend on s. In this case, the admissible growth in practise does not cause any great difficulty, as practically all functions encountered in the applications have exponential growth (considerably weaker than (2.8)). The restriction of the domain of definition of A_z to simple functions does not diminish the possibility of interpolation, as the simple functions constitute a dense subset in $L_p(\mathbb{T}^N)$.

Let us turn to the inequalities (2.6) and (2.7). Let us fix an arbitrary $\varepsilon > 0$ and set $s(z) = (\frac{N-1}{2} + \varepsilon) z$. Then the operators $E_\lambda^{s(t)}$ satisfy all the assumptions of Stein's interpolation theorem with $p_1 = 1$ and $p_0 = 2$, and therefore for $0 < t < 1$

$$\|E_\lambda^{s(t)}\|_{L_p(\mathbb{T}^N)} \le \text{const} \|f\|_{L_p(\mathbb{T}^N)},$$

where

$$s(t) = \left(\frac{N-1}{2} + \varepsilon\right) t, \quad \frac{1}{p} = \frac{1-t}{2} + \frac{t}{1},$$

whence eliminating t

$$s = (N - 1 + 2\varepsilon)\left(\frac{1}{p} - \frac{1}{2}\right), \quad 1 < p < 2.$$

Thus, if $1 \le p \le 2$ the Riesz means of order s of the Fourier series of any function $f \in L_p(\mathbb{T}^N)$ converge for $s > (N-1)(\frac{1}{p} - \frac{1}{2})$ to f in the $L_p(\mathbb{T}^N)$-metric (cf. Stein (1958)). By duality, an analogous statement is true for $p > 2$ with $s > (N-1)(\frac{1}{2} - \frac{1}{p})$, or for any p with

$$s > (N-1)\left|\frac{1}{p} - \frac{1}{2}\right|, \quad p \ge 1. \tag{2.9}$$

How sharp is condition (2.9)? The divergence result, which we are now going to state, joins up with condition (2.9) the closer p comes to 1 or ∞ (Babenko (1973a)): if p does not belong to the interval $I_N = [\frac{2N}{N+1}, \frac{2N}{N-1}]$ (i. e. if $|\frac{1}{p} - \frac{1}{2}| > \frac{1}{2N}$) there exists a function $f \in L_p(\mathbb{T}^N)$ whose Riesz mean $E_\lambda^s f$ for

$$0 \le s \le N\left|\frac{1}{p} - \frac{1}{2}\right| - \frac{1}{2}. \tag{2.10}$$

do not converge to f in the $L_p(\mathbb{T}^N)$-metric. In other words, under the hypothesis (2.10) the multiple trigonometric system is not a basis for Riesz summability in $L_p(\mathbb{T}^N)$. Let us show how this result can be obtained from Il'in's estimate (1.19). To this end let us introduce the extremely important *zeta function*

$$\zeta_\tau(x) = \sum_{0 \ne n \in \mathbb{Z}^n} |n|^{-\tau} e^{inx}. \tag{2.11}$$

As the series in the right hand side may diverge in the L_p-metric and also almost everywhere, it is necessary to make precise in what sense we take the equality (2.11). We will interpret this equality so that to the right stands the Fourier series of ζ_τ. In other words, $\zeta_\tau(x)$ is a function, orthogonal to unity, whose Fourier coefficients with respect to the multiple trigonometric system equal $|n|^{-\tau}$:

$$(2\pi)^{-N} \int_{\mathbb{T}^N} \zeta_\tau(x) e^{inx} dx = |n|^{-\tau}, \quad n \ne 0. \tag{2.12}$$

The first question which arises from such a definition is the question of existence of $\zeta_\tau(x)$. If $\tau > N/2$ then this follows from the convegence of the series $\sum_{n \ne 0} |n|^{-2\tau}$ along with the Riesz-Fischer theorem. The proof of the existence of $\zeta_\tau(x)$ in the case $0 < \tau \le N/2$ was first obtained by V. A. Il'in (cf. (1958)). In this paper a detailed description of the singularities of $\zeta_\tau(x)$ for all $\tau > 0$ were given and, in particular, it was proved that

$$\zeta_\tau(x) = c_\tau |x|^{\tau - N} + \beta_\tau(x), \quad 0 < \tau < N, \tag{2.13}$$

with $\beta_\tau(x) \in C(\mathbb{T}^N)$ infinitely differentiable in the interior of $\mathbb{T}^N$. The integral operator with kernel $\zeta_\tau(x-y)$ coincides with the fractional power of the Laplace operator $(-\Delta)^{-\tau/2}$ on functions orthogonal to unity (i. e. functions orthogonal to the kernel of Laplace operator on $\mathbb{T}^N$). Therefore ζ_τ is sometimes called the *kernel of fractional order*.

If $D_\mu(x)$ is the Dirichlet kernel for spherical summation:

$$D_\mu(x) = \sum_{|n|\le\mu} e^{inx},$$

then the spherical Riesz means of the series (2.11) can be written as

$$R_\mu^s\zeta_\tau(x) = \int_1^\mu \left(1 - \frac{t^2}{\mu^2}\right)^s t^{-\tau} dD_t(x). \tag{2.14}$$

Comparing them with the means of the Dirichlet kernel

$$D_\mu^s(x) = \int_1^\mu \left(1 - \frac{t^2}{\mu^2}\right)^s dD_t(x),$$

from which (2.14) differs only by the factor τ^{-t} in the integral, it is natural to conjecture (this is so in the case of interest to us) that

$$D_\mu^s(x) \sim \mu^\tau R_\mu^s\zeta_\tau(x). \tag{2.15}$$

Let $1 \le p < \frac{2N}{N+1}$. If s satisfies the condition

$$0 \le s < N\left(\frac{1}{p} - \frac{1}{2}\right) - \frac{1}{2},$$

then one can always choose τ such that

$$N - \frac{N}{p} < \tau < \frac{N-1}{2} - s.$$

In view of (2.13) the left hand side of this inequality guarantees that ζ_τ belongs to $L_p(\mathbb{T}^N)$, while the inequality to the right, together with the estimate (1.19) and relation (2.15), shows that the norm of $R_\mu^s\zeta_\tau$ in $L_p(\mathbb{T}^N)$ cannot be bounded. The limiting case $s = N(\frac{1}{p} - \frac{1}{2}) - \frac{1}{2}$ requires somewhat more tedious considerations.

The "necessary" condition (2.10) so far known and the sufficient one (2.9), for convergence in $L_p(\mathbb{T}^N)$, remain in force also for partial sums defined by elliptic polynomials and their Riesz means (Hörmander (1969)).

A comparison of (2.9) and (2.10) reveals that they deviate from each other. The Carleson-Sjölin theorem and also the theorem of Fefferman and V. Z. Meshkov, to be set forth in the following section, give reasons to believe that the "necessary" conditions (2.10) are sharp and that this deviation comes from the circumstance that the sufficient conditions are not complete.

§2. Convergence of Fourier Integral Expansions in L_p

2.1. The Case $N = 1$. The convergence of one-dimensional Fourier expansions in the metric $L_p(\mathbb{R}^1)$ is a consequence of the boundedness in L_p of the *Hilbert transform*

$$Hf(x) = \frac{1}{\pi}\int_{-\infty}^{\infty} f(x-t)\frac{dt}{t} \tag{2.16}$$

(the integral is taken in the principal value sense). It is not hard to see that

$$\mathcal{F}(Hf)(x) = \widehat{Hf}(\xi) = (i \operatorname{sign} \xi) \cdot \hat{f}(\xi), \tag{2.17}$$

and, consequently, $\varphi(\xi) = \operatorname{sign}\xi$ is a multiplier from L_p into L_p. Let us recall the definition of a multiplier in general. A function $\varphi(\xi)$, $\xi \in \mathbb{R}^N$, is called a *multiplier* from $L_p(\mathbb{R}^N)$ into $L_q(\mathbb{R}^N)$ if the operator T_φ, defined by the equality

$$\mathcal{F}(T_\varphi f)(\xi) = \widehat{T_\varphi f}(\xi) = \varphi(\xi)\hat{f}(\xi), \tag{2.18}$$

is of *strong type* (p, q), i. e. one has the inequality

$$\|T_\varphi f\|_{L_q(\mathbb{R}^N)} \le \text{const} \cdot \|f\|_{L_p(\mathbb{R}^N)}. \tag{2.19}$$

The *class of multipliers* from $L_p(\mathbb{R}^N)$ into $L_p(\mathbb{R}^N)$ is of great interest and will be denoted by the symbol $M_p(\mathbb{R}^N)$. It is clear that if $\varphi \in M_p$ and $\psi \in M_p$ then $\varphi + \psi \in M_p$ and $\varphi \cdot \psi \in M_p$. If $\varphi \equiv 1$ then T_φ is the identity operator so that $1 \in M_p$. Moreover, if $\psi(\xi) = \varphi(\xi + h)$ then

$$T_\psi f(x) = e^{-ixh}T_\varphi[e^{ixh}f(x)],$$

so that $\varphi(\xi) \in M_p$ entails that $\varphi(\xi + h) \in M_p$.

In the case $N = 1$ the operator (2.16) is bounded from $L_p(\mathbb{R}^1)$ into $L_p(\mathbb{R}^1)$ so that (2.17) implies that $\operatorname{sign}\xi \in M_p(\mathbb{R}^1)$. Then $h(\xi) = \frac{1}{2}(1 + \operatorname{sign}\xi)$, i. e.

$$h(\xi) = \begin{cases} 1, & \xi > 0, \\ 0, & \xi < 0, \end{cases}$$

is likewise a multiplier in L_p (i. e. from L_p into L_p). From this follows that the characteristic function of any halfaxis, and thus also any segment on the axis $\mathbb{R}^1$ is a multiplier in L_p, when $1 < p < \infty$.

Obviously the norm of the operator

$$S_\mu f(x) = \frac{1}{2\pi}\int_{-\mu}^{\mu} \hat{f}(\xi)e^{i\xi x}d\xi \tag{2.20}$$

from L_p into L_p does not depend on μ. Therefore the boundedness (and so the continuity) in L_p of the partial sums of Fourier integrals holds if and only if the characteristic function of the interval $\{\xi : |\xi| \leq 1\}$ is a multiplier in L_p, i. e. $1 < p < \infty$.

For $p = 1$ the situation deteriorates, since the operators (2.20) are not bounded from $L_1(\mathbb{R}^1)$ into $L_1(\mathbb{R}^1)$. Recall that this follows by duality from the unboundedness of the operator (2.20) from $C(\mathbb{R}^1)$ into $C(\mathbb{R}^1)$, i. e. the fact that there exists a continuous function with an unboundedly divergent Fourier integral.

Thus, the convergence of the Fourier integral expansion of a function $f \in L_p(\mathbb{R}^1)$ to itself in the L_p-metric is guaranteed only for $1 < p < \infty$.

2.2. Bases and the Problem of Spherical Multipliers. Passing to dimensions $N > 1$, let us turn our attention to the following intuitively clear fact: if $\varphi(\xi)$, $\xi \in \mathbb{R}^N$, does not depend on the variable ξ_k then the operator T_φ, defined by formula (2.18), is the identity in this variable, i. e. acts effectively only in the remaining variables. From this follows a more precise statement: if $\varphi \in M_p(\mathbb{R}^{N-1})$ then the same object φ, considered as a function in N variables, belongs to $M_p(\mathbb{R}^N)$.

In particular, it follows from the properties of the Hilbert transform that the characteristic function of a halfplane in $\mathbb{R}^2$ is a multiplier in $L_p(\mathbb{R}^2)$, $1 < p < \infty$. Taking the product of such multipliers we see that the characteristic function of any rectangle $\{a \leq x_1 \leq b, c \leq x_2 \leq d\}$ is a multiplier. Exactly in the same way we get that the characteristic function of any polyhedron in $\mathbb{R}^N$ is a multiplier in $L_p(\mathbb{R}^N)$, $1 < p < \infty$.

Thus, the partial Fourier integrals of a function in $L_p(\mathbb{R}^N)$, taken with respect of the dilations of a polyhedron, converge in the metrić of $L_p(\mathbb{R}^N)$, $1 < p < \infty$.

Is the same property true for spherical domains, i. e. is the characteristic function of the unit ball $B = \{\xi \in \mathbb{R}^N : |\xi| \leq 1\}$ a multiplier in $L_p(\mathbb{R}^N)$, $1 < p < \infty$? If we do not put any restrictions to p then the answer to this question is negative. Set

$$Tf(x) = (2\pi)^{-N/2} \int_{|\xi|\leq 1} \hat{f}(\xi) e^{-ix\xi} d\xi. \tag{2.21}$$

The operator (2.21) is an integral operator whose kernel is given by formula (29), so that

$$Tf(x) = (2\pi)^{-N/2} \int_{\mathbb{R}^N} f(y) J_{\frac{N}{2}}(|x-y|)|x-y|^{-\frac{N}{2}} dy. \tag{2.22}$$

If we take for f the characteristic function of a ball of sufficiently small radius (so that inside this ball the Bessel function does not oscillate), then the function Tf obtained cannot, in view of the slow decrease at infinity, belong to $L_p(\mathbb{R}^N)$ for $|\frac{1}{p} - \frac{1}{2}| \geq \frac{1}{2N}$ (this follows readily from the asymptotics of the Bessel function).

The question if the operator (2.21) is bounded from $L_p(\mathbb{R}^N)$ into $L_p(\mathbb{R}^N)$ for $|\frac{1}{p} - \frac{1}{2}| < \frac{1}{2N}$ is known as the *multiplier problem for the ball*, which for a long time was open. Of course, in the trivial case $p = 2$ the answer to this question is positive. However, it was a surprise that only this case exhausts the set of p's for which (2.21) is bounded in $L_p(\mathbb{R}^N)$, $N \geq 2$.

Theorem (Fefferman (1970)). *The characteristic function of the unit ball is a Fourier multiplier in* $L_p(\mathbb{R}^N)$ *if and only if* $p = 2$.

At this juncture we interrupt our presentation and turn to a problem, at the first glance very remote from our subject, namely the *Kakeya problem* which can be formulated as follows. Suppose that in the plane there lies a needle covered by paint. It is required, without lifting it from the plane, to turn the needle 360° moving it in such a manner that the part which gets colored by the needle has the least area. The determination of this minimal area is precisely the Kakeya problem.

In the first (erroneous) solutions of this problem there appeared as minimal domains colored domains lying in the interior of a curve of asteroid type. However the correct answer, due to Besicovich, contained different ones: by complicating the movement of the needle one can make the colored area arbitrarily small. At the basis of Besicovich's construction are "needleshaped" domains, constructed as follows. The interval $[0, 1]$ on the abscissa axis is divided into 2^k equal parts, each taken as the basis of a triangle. The heights of the triangles (and, thus, likewise their areas) are the same, but the vertices are situated in inverse order: the more to the right the basis of the triangle is, the more to the left one takes its vertex. When the heights for $k \to \infty$ grow as $\log k$, it is clear that the sum of the areas of the triangles grows as $\log k$. Therefore it is natural to expect that the set E equal to the union of the triangles likewise grows as $\log k$. However, it is remarkable that one can position the vertices in such a way that the measure of E stays bounded (and, in particular, does not exceed 17). This means that the triangles intersect many times and have a large common portion.

Precisely this circumstance was at the basis of the construction of Fefferman's counter-example. If one prolongs the sides of each triangle below, under the abscissa axis, the corresponding prolongations of the triangles (which follows in a straightforward manner from the position of their vertices) do not intersect pairwise. Moreover, one can find a function whose support is in the interior of this prolongation. A computation reveals that the spherical partial Fourier integrals of this function accumulate on the set E and that their norm in $L_p(E)$, when k gets large, grows faster than the $L_p(\mathbb{R}^2)$-norm of the function itself.

The fact that Kakeya's problem, which is so remote from harmonic analysis, proved to be the decisive step in the final solution of the multiplier problem for the ball (however not in the expected way) leads to speculations and to far reaching considerations. Let us limit ourselves just to state some consequences of the properties of the above construction. An analysis reveals that there exist no multipliers in $L_p(\mathbb{R}^2)$, $p \neq 2$, which are characteristic functions of bounded convex sets bounded by, say, the level lines of an elliptic polynomial. Moreover, if the boundary of a convex set is a piecewise smooth curve and its characteristic function is a multiplier in $L_p(\mathbb{R}^2)$, $p \neq 2$, then it must be a polygone.

From the negative solution of the multiplier problem for the ball in $\mathbb{R}^2$ follows also its negative solution for $N > 2$ (which is obtained by "stretching" the ball along the corresponding coordinate axies). Furthermore, as was proved in the last section, it follows from here that the multiple trigonometric system is not a basis in $L_p(\mathbb{T}^N)$, $p \neq 2$, for spherical summation. Finally, in the following chapter, as well as from this result, one can get a function in $L_p(\mathbb{T}^N)$, $p \neq 2$, with a Fourier series which is divergent on a set of positive measure for spherical summation.

2.3. Bases of Riesz Means. The result on divergence in $L_p(\mathbb{R}^N)$ forces one to return once more to the Riesz means of spherical partial integrals. By standard arguments one can reduce the summability problem in $L_p(\mathbb{R}^N)$ to the following: for which s is the function

$$\rho_s(\xi) = \begin{cases} (1-|\xi|^2)^s, & |\xi| < 1, \\ 0, & |\xi| \geq 1 \end{cases} \tag{2.23}$$

a multiplier in $L_p(\mathbb{R}^N)$?

It is not hard to see that the corresponding operator

$$R^s f(x) = (2\pi)^{-N} \int_{|\xi|\leq 1} (1-|\xi|^2)^s \hat{f}(\xi) e^{ix\xi} d\xi \tag{2.24}$$

is an integral operator with the kernel in (47) taken with $\mu = 1$. Therefore,

$$R^s f(x) = \text{const} \int_{\mathbb{R}^N} |y|^{-\frac{N}{2}-s} J_{\frac{N}{2}+s}(|y|) f(x+y) dy. \tag{2.25}$$

Taking for f the characteristic function of a ball of sufficiently small radius it follows immediately from the asymptotics of the Bessel function and from (2.25) that $R^s f(x)$ decreases at infinity as $|x|^{-s-\frac{N+1}{2}}$. Consequently, $R^s f$ does not belong to $L_p(\mathbb{R}^N)$ provided

$$1 \leq p \leq \frac{N}{\frac{N+1}{2} + s}.$$

Otherwise put, if $1 \leq p \leq \frac{N}{N+1}$ and

$$0 \leq s \leq N\left(\frac{1}{p} - \frac{1}{2}\right) - \frac{1}{2},$$

then there exists a function in $L_p(\mathbb{R}^N)$ with divergent Riesz means $R^s f(x)$ in $L_p(\mathbb{R}^N)$. By duality this statement is true for

$$p \notin I_N = \left[\frac{2N}{N+1}, \frac{2N}{N-1}\right], \quad 0 \le s \le N\left|\frac{1}{p} - \frac{1}{2}\right| - \frac{1}{2}. \tag{2.26}$$

Hence, it is too much to hope that the summability problem in $L_p(\mathbb{R}^N)$ can be solved for Riesz means of small order (independent of N) for $p \notin I_N$. In the same way as in the case of multiple series (cf. Sect. 1.3) one can obtain results on convergence by interpolating between the L_1 and the L_2-estimates, and then the sufficient conditions for convergence of $R^s_\mu f$ take the form

$$s > (N-1)\left|\frac{1}{p} - \frac{1}{2}\right|, \quad p \ge 1. \tag{2.27}$$

Thus, the function (2.23) is a multiplier in $L_p(\mathbb{R}^N)$ if (2.27) is fulfilled and is not a multiplier if (2.26) holds. As one sees, between the conditions (2.26) and (2.27) there is a "margin" for $p \notin I_N$, in size equal to

$$\frac{1}{2} - \left|\frac{1}{p} - \frac{1}{2}\right| > 0.$$

In the two-dimensional case one has been able to eliminate this margin thanks to the following remarkable result.

Theorem (Carleson-Sjölin (1972)). *The function* (2.23) *is a multiplier in* $L_p(\mathbb{R}^2)$ *for* $p \in I_2 = [\frac{4}{3}, 4]$ *and any* $s > 0$.

The proof of this theorem is hard. Let us just remark that it suffices to prove it for $p = 4$. The result for $p = \frac{4}{3}$ then follows by duality ($\frac{1}{4} + \frac{3}{4} = 1$) and the remaining cases follow by interpolation.

If $p \notin I_2$, i. e. if $|\frac{1}{p} - \frac{1}{2}| \ge 4$, it follows by interpolation from the Carleson-Sjölin theorem and condition (2.27), that the function (2.23) is a Fourier multiplier in $L_p(\mathbb{R}^2)$ provided

$$s > 2\left|\frac{1}{p} - \frac{1}{2}\right| - \frac{1}{2}, \quad p \notin I_2.$$

Thus, the theorems of Fefferman and Carleson-Sjölin give an exhaustive solution to the question of the convergence in $L_p(\mathbb{R}^2)$ of spherical Riesz means of Fourier integral expansions.

It is natural to hope that the Carleson-Sjölin theorem extends also to the case $N > 2$, i. e. that the function (2.23) for every $s > 0$ is a multiplier in $L_p(\mathbb{R}^N)$, $p \in I_N$. However, at the present moment this has not been proved.

Assuming that the N-dimensional analogue of the Carleson-Sjölin theorem is true, it would follow from Stein's interpolation theorem that the function (2.23) is a multiplier in $L_p(\mathbb{R}^N)$ provided $p \notin I_N$ and

$$s > N\left|\frac{1}{p} - \frac{1}{2}\right| - \frac{1}{2}. \tag{2.28}$$

For certain $p \notin I_N$ this result has in fact been verified. Namely, Fefferman (1970) has proved that condition (2.28) is sufficient for the function (2.23) to be a multiplier in $L_p(\mathbb{R}^N)$ for $|\frac{1}{p} - \frac{1}{2}| > \frac{N+1}{4N}$. Refining this proof, V. Z. Meshkov (1978) extended the validity of condition (2.28) to p's satisfying the inequality

$$\left|\frac{1}{p} - \frac{1}{2}\right| > \frac{1}{N+1}.$$

This result serves as a supplementary argument to the belief that condition (2.28), apparently, guarantees the convergence of the means $R_\mu^s f$ in L_p for all $p \notin I_N$.

§3. Multipliers

3.1. The Theorem of Marcinkiewicz. Let us recall that in Sect. 2.1 we termed a function $\varphi(\xi)$ a *multiplier* from $L_p(\mathbb{R}^N)$ into $L_q(\mathbb{R}^N)$ if the Fourier transform of any function in $L_p(\mathbb{R}^N)$ becomes the Fourier transform of some function in $L_q(\mathbb{R}^N)$ after multiplication by $\varphi(\xi)$. One can likewise introduce an analogous notion in the periodic case.

A sequence of complex numbers $\lambda(n)$, $n \in \mathbb{Z}^N$, is called a *multiplier* from $L_p(\mathbb{T}^N)$ into $L_q(\mathbb{T}^N)$ if the operator

$$T_\lambda f(x) = \sum_{n \in \mathbb{Z}^N} \lambda(n) f_n e^{inx} \tag{2.29}$$

is of strong type (p, q), i. e. if

$$\|T_\lambda\|_{L_q(\mathbb{T}^N)} \leq \text{const} \cdot \|f\|_{L_p(\mathbb{T}^N)}. \tag{2.30}$$

The norm of the operator T_λ is called the *multiplier norm* of λ.

In the most important case $p = q$ multipliers from $L_p(\mathbb{T}^N)$ into $L_p(\mathbb{T}^N)$ will simply be referred to as multipliers on $L_p(\mathbb{T}^N)$ and the set of all such multipliers will be denoted $M_p(\mathbb{T}^N)$. The following example illustrates the importance of multipliers.

Let us consider the *Poisson equation*

$$\Delta u(x) \equiv \sum_{k=1}^{N} \frac{\partial^2 u}{\partial x_k^2} = f(x), \quad x \in \mathbb{T}^N,$$

in periodic boundary conditions. A solution which is orthogonal to unity can be obtained by the method of separation of variables:

$$u(x) = \sum_{0 \neq n \in \mathbb{Z}^N} \frac{1}{|n|^2} f_n e^{inx}.$$

A major problem in the theory of elliptic partial differential equations (to which Poisson's equation belongs) is the estimating of the second derivatives

$$\frac{\partial^2 u}{\partial x_j \partial x_k} = \sum_{0 \neq n \in \mathbf{Z}^N} \frac{n_j n_k}{|n|^2} f_n e^{inx}, \quad j,k = 1,\ldots,N. \tag{2.31}$$

If $\lambda(n) = n_j n_k/|n|^2$ is a multiplier on $L_p(\mathbb{T}^N)$ (that this is in fact so follows from Marcinkiewicz's theorem *infra*) then from (2.31) we get at once the estimate

$$\left\| \frac{\partial^2 u}{\partial x_j \partial x_k} \right\|_{L_p(\mathbb{T}^N)} \leq \text{const}\, \|f\|_{L_p(\mathbb{T}^N)}, \quad 1 < p < \infty, \quad j,k = 1,\ldots,N,$$

which yields, in view of the definition (50) of the norm in the Sobolev classes W_p^l,

$$\|u\|_{W_p^2(\mathbb{T}^N)} \leq \text{const}\, \|f\|_{L_p(\mathbb{T}^N)}, \quad 1 < p < \infty.$$

Among the classes $M_p(\mathbb{T}^N)$ it is $M_2(\mathbb{T}^N)$, i. e. the class of multipliers on $L_2(\mathbb{T}^N)$, that has the simplest structure. Indeed, applying to (2.29) Parseval's formula we find

$$\|T_\lambda\|^2_{L_2(\mathbb{T}^N)} = (2\pi)^{-N} \sum |\lambda(n)|^2 |f_n|^2$$

and so if $\lambda(n) \in M_2(\mathbb{T}^N)$, then in view of (2.30), for each $f \in L_2(\mathbb{T}^N)$ we must have

$$\sum |\lambda(n)|^2 |f_n|^2 \leq \text{const} \sum |f_n|^2.$$

Clearly this holds if and only if $\lambda(n)$ is a bounded sequence.

The case $p \neq 2$ is considerably harder. Let us remark that if $\lambda \in M_p(\mathbb{T}^N)$ then $\overline{\lambda} \in M_p(\mathbb{T}^N)$ so that by duality $\lambda \in M_q(\mathbb{T}^N)$, where $\frac{1}{p} + \frac{1}{q} = 1$.

Upon applying the classical Riesz interpolation theorem we find that $\lambda \in M_p(\mathbb{T}^N)$ for any r beween p and q, and in particular for $r = 2$. Therefore, by what has been said, $|\lambda(n)| \leq$ const, i. e. each multiplier in $L_p(\mathbb{T}^N)$ is bounded, but in contrast to the case $p = 2$ the converse is not true.

A sufficient condition for $\lambda \in M_p(\mathbb{T}^N)$ was found by Marcinkiewicz. For the formulation of this result we introduce the notation

$$\Delta_j \lambda(n) = \lambda(n + e^j) - \lambda(n),$$

where e^j is the vector with components $e^j_k = \delta_{jk}$. If α is a multi-index, we set

$$\Delta^\alpha \lambda(n) = \Delta_1^{\alpha_1} \ldots \Delta_N^{\alpha_N} \lambda(n).$$

For arbitrary multi-indices k and α we write

$$(k, \alpha) = \{n \in \mathbb{Z}^N : 2^{k_j} \leq |n_j| \leq 2^{k_j + \alpha_j}, \quad j = 1,\ldots,N\}.$$

Set further $I = \sum_{j=1}^{N} e^j = (1, 1, \ldots, 1)$. As usual, the inequality $\alpha \geq \beta$ means that $\alpha_j \geq \beta_j$ for all $j = 1, \ldots, N$.

Theorem (Marcinkiewicz (1939)). *Assume that the bounded sequence $\lambda(n)$ satisfies for each α, $0 \leq \alpha \leq I$, and each $k \geq 0$ the inequality*

$$\sum_{n \in (k,\alpha)} |\Delta^\alpha \lambda(n)| \leq \text{const}, \tag{2.32}$$

where the constant does not depend on k. Then $\lambda(n)$ is a multiplier on $L_p(\mathbb{T}^N)$, $1 < p < \infty$.

3.2. The Nonperiodic Case. Marcinkiewicz's theorem has turned out to be rather fruitful also in the nonperiodic case, i. e. in the study of the Fourier transform. Using this theorem one has been able to obtain sufficient conditions, of considerable interest for the applications, for the membership of a bounded function $\varphi(\xi)$ in the class $M_p(\mathbb{R}^N)$, $1 < p < \infty$ (cf. Lizorkin (1963), Mikhlin (1956), Hörmander (1960)). Let us state the most useful variant for our purposes of these conditions due to P. I. Lizorkin.

Theorem (Lizorkin (1963)). *Assume that the function $\varphi(\xi)$ for each multi-index α, $0 \leq \alpha \leq I$, has a continuous derivative $D^\alpha \varphi(\xi)$ at points $\xi \in \mathbb{R}^N$ with $\xi_j \neq 0$ if $\alpha_j = 1$ satisfying the condition*

$$|\xi^\alpha D^\alpha \varphi(\xi)| \leq \text{const}, \tag{2.33}$$

where the constant does not depend on ξ. Then $\varphi(\xi)$ is a multiplier in $L_p(\mathbb{R}^N)$, $1 < p < \infty$.

Note that (2.33) for $\alpha = 0$ implies the boundedness of φ in $\mathbb{R}^N$. This is not a coincidence, as precisely as in the periodic case it is not hard to prove that $M_p(\mathbb{R}^N) = M_q(\mathbb{R}^N)$, where $\frac{1}{p} + \frac{1}{q} = 1$, and by the M. Riesz interpolation theorem, $M_p(\mathbb{R}^N) \subset M_r(\mathbb{R}^N)$ for any r between p and q. In particular, if $\varphi \in M_p(\mathbb{R}^N)$ then $\varphi \in M_2(\mathbb{R}^N)$ so that

$$\int_{\mathbb{R}^N} |\varphi(\xi)|^2 |\hat{f}(\xi)|^2 d\xi \leq \text{const} \int_{\mathbb{R}^N} |\hat{f}(\xi)|^2 d\xi,$$

which again implies that $\varphi(\xi)$ is bounded.

Note however that if $\varphi(\xi)$ is a multiplier from $L_p(\mathbb{R}^N)$ into $L_q(\mathbb{R}^N)$, where $p < q$, then $\varphi(\xi)$ need not necessarily be bounded. Indeed, let us show that for $1 < p < q < \infty$ the function

$$\varphi(\xi) = |\xi|^{-\tau}, \quad \tau = N\left(\frac{1}{p} - \frac{1}{q}\right) \tag{2.34}$$

is a multiplier from $L_p(\mathbb{R}^N)$ into $L_q(\mathbb{R}^N)$. To this end we consider the function (the integral is taken in the principal value sense)

$$K_\tau(x) = \int_{\mathbb{R}^N} |\xi|^{-\tau} e^{ix\xi} d\xi. \tag{2.35}$$

Setting $\theta = x/|x|$ and making a change of variable, we get

$$K_\tau(x) = |x|^{\tau - N} \int_{\mathbb{R}^N} |\xi|^{-\tau} e^{i\theta\xi} d\xi.$$

In view of the spherical symmetry of factor in front of the exponential, the integral must be independent of θ so that

$$K_\tau(x) = \text{const} \cdot |x|^{\tau - N}. \tag{2.36}$$

The convolution of the kernel (2.36) with a function $f \in L_p(\mathbb{R}^N)$, defined as

$$K_\tau * f(x) = \int_{\mathbb{R}^N} K_\tau(x-y) f(y) dy,$$

is estimated with the aid of *Sobolev's lemma* (Sobolev (1950)), which is the N-dimensional analogue of a well-known theorem by Hardy-Littlewood:

$$\|K_\tau * f\|_{L_q(\mathbb{R}^N)} \le \text{const} \cdot \|f\|_{L_p(\mathbb{R}^N)}, \quad \tau = N\left(\frac{1}{p} - \frac{1}{q}\right). \tag{2.37}$$

It remains to remark that, in view of the properties of the Fourier transform and inequality (2.25), this convolution equals

$$K_\tau * f(x) = (2\pi)^{N/2} \int_{\mathbb{R}^N} \hat{f}(\xi) |\xi|^{-\tau} e^{i\theta\xi} d\xi.$$

The last equality together with the estimate (2.37) implies that the function (2.34) is a multiplier from $L_p(\mathbb{R}^N)$ into $L_q(\mathbb{R}^N)$.

Now we interrupt our discussion on the theory of multipliers in $\mathbb{R}^N$ and turn to the periodic case. The obvious analogue of the multiplier (2.34) is the sequence

$$\lambda(n) = |n|^{-\tau}, \quad \tau = N\left(\frac{1}{p} - \frac{1}{q}\right) > 0 \tag{2.38}$$

and then to the kernel (2.35) corresponds the zeta function (cf. (2.11))

$$\zeta_\tau(x) = \sum_{0 \ne n \in \mathbb{Z}^n} |n|^{-\tau} e^{inx}.$$

In view of (2.13) the zeta function $\zeta_\tau(x)$ has exactly the same properties as the kernel $K_\tau(x)$ and therefore the sequence (2.38) is a multiplier from $L_p(\mathbb{R}^N)$ into $L_q(\mathbb{R}^N)$.

But let us turn anew to multipliers in $\mathbb{R}^N$. The generalization of Marcinkiewicz's theorem formulated above allows us, as we have already recorded, to conclude in a very simple manner that a given function is a multiplier in $L_p(\mathbb{R}^N)$. Let us illustrate this at the hand of the following example:

$$\psi(\xi) = |\xi|^\tau (1 + |\xi|^2)^{-\tau/2}. \tag{2.39}$$

If we write $b(t) = [t(1+t^2)^{-1/2}]^\tau$ then $\psi(\xi) = b(|\xi|)$. It is not hard to see that

$$|b^{(k)}(t)| \leq \text{const} \cdot t^{-k}, \quad t > 0, \quad k = 1, 2, \dots,$$

and this shows at once that the function (2.39) satisfies condition (2.33) and therefore must be a multiplier in $L_p(\mathbb{R}^N)$, $1 < p < \infty$.

Another important example is provided by the functions

$$\psi_k(\xi) = \operatorname{sign} \xi_k, \quad k = 1, 2, \dots, N.$$

Each of these functions trivially satisfies inequality (2.33) and is a multiplier on $L_p(\mathbb{R}^N)$, $1 < p < \infty$ (note that this is not the case for $p = 1$). In particular, this implies the boundedness of the Hilbert transform in L_p (cf. Sect. 2 of this chapter).

Likewise the functions

$$\varphi_k(\xi) = \xi_k/|\xi|, \quad k = 1, 2, \dots, N,$$

are multipliers on $L_p(\mathbb{R}^N)$.

Let us have a look at yet another important function:

$$\varphi_\tau(\xi) = (1+|\xi|^2)^{-\tau/2}, \quad k = 1, 2, \dots, N, \quad \tau = N\left(\frac{1}{p} - \frac{1}{q}\right) > 0. \qquad (2.40)$$

It follows from the properties of the multipliers (2.34) and (2.39) that φ_τ is a multiplier from $L_p(\mathbb{R}^N)$ into $L_q(\mathbb{R}^N)$.

Let us observe that T_{φ_τ}, in view of the definition of the Liouville classes (cf. Sect. 10 of the introduction), maps $L_p(\mathbb{R}^N)$ into $L_p^l(\mathbb{R}^N)$. This leads to the idea that it might be possible to obtain imbedding theorems for Liouville classes from the multiplier properties of the function (2.40). Let us prove, on the basis of these properties, the following imbedding theorem (cf. Lizorkin (1963)):

$$L_p^l(\mathbb{R}^N) \to L_q^m(\mathbb{R}^N), \quad l - \tfrac{N}{p} = m - \tfrac{N}{q} \qquad (2.41)$$

(here $l > m > 0$, $1 < p < q < \infty$); no immediate applications of this will be made here.

Setting $\tau = l - m > 0$ we can write

$$(1+|\xi|^2)^{-l/2} = (1+|\xi|^2)^{-m/2} \cdot (1+|\xi|^2)^{-\tau/2}.$$

The multiplier (2.40) maps $L_p(\mathbb{R}^N)$ into $L_q(\mathbb{R}^N)$, while the multiplier $(1+|\xi|^2)^{-m/2}$ takes $L_q(\mathbb{R}^N)$ into the Liouville class $L_q^m(\mathbb{R}^N)$. Since their composition then maps $L_p(\mathbb{R}^N)$ into $L_p^l(\mathbb{R}^N)$, the required imbedding (2.41) follows.

3.3. Are Multipliers in $\mathbb{R}^N$ Multipliers on $\mathbb{T}^N$? In the preceding section we had several occasions to turn the attention to the analogy between multipliers in $\mathbb{R}^N$ and in $\mathbb{T}^N$. That this analogy is not a coincidence is established in the following theorem due to de Leeuw.

Theorem (de Leeuw (1965)). *Let $\lambda(\xi)$ be a function which is continuous at points of the integer lattice $\mathbb{Z}^N$ and assume that it is a multiplier in $L_p(\mathbb{R}^N)$, $1 \le p \le \infty$. Then the sequence $\lambda(n)$ is a multiplier on $L_p(\mathbb{T}^N)$.*

Let us remark that the assumption of the continuity of $\lambda(\xi)$ at points of $\mathbb{Z}^N$ is necessary, because if we change a function in $M_p(\mathbb{R}^N)$ in a set of measure zero its multiplier properties are not affected. In order to illustrate the power of de Leeuw's theorem let us consider the function (2.39). As $\psi \in M_p(\mathbb{R}^N)$, $1 < p < \infty$, then by this theorem

$$\lambda(n) = |n|^\tau \cdot (1 + |n|^2)^{-\tau/2}$$

is a multiplier in $L_p(\mathbb{T}^N)$. Taking into account that the sequence (2.38) is a multiplier from $L_p(\mathbb{T}^N)$ into $L_q(\mathbb{T}^N)$ we conclude that

$$\lambda_\tau(n) = (1 + |n|^2)^{-\tau/2}, \quad \tau = N\left(\frac{1}{p} - \frac{1}{q}\right)$$

is a multiplier from $L_p(\mathbb{T}^N)$ into $L_q(\mathbb{T}^N)$. From this statement, similarly to how was done in Sect. 2.3, it is not hard to derive an imbedding theorem for Liouville classes:

$$L_p^l(\mathbb{T}^N) \to L_q^m(\mathbb{T}^N), \quad l - \tfrac{N}{p} = m - \tfrac{N}{q},$$

where $l > m > 0$, $1 < p < q < \infty$.

With the aid of de Leeuw's theorem it is not hard to show that the sequences

$$\lambda_j(n) = \frac{n_j}{|n|}, \quad j = 1, 2, \ldots, N,$$

are multipliers in $L_p(\mathbb{T}^N)$ and therefore also all products $\lambda_j(n) \cdot \lambda_k(n)$. From this follows, in particular, the estimate of the second derivatives of the solutions of the Poisson equation in terms of the right hand member (cf. (2.31)).

De Leeuw's theorem allows to translate multipliers in $L_p(\mathbb{R}^N)$ into multipliers in $L_p(\mathbb{T}^N)$. There arises the question whether the converse is true. Namely, if the function $\mathbb{Z}^N$ is continuous at the points of $\mathbb{Z}^N$ and if $\lambda(n)$ is a multiplier on $L_p(\mathbb{T}^N)$, does it follow that $\lambda(\xi) \in M_p(\mathbb{R}^N)$? The following simple example gives a negative answer to this question:

$$\lambda(\xi) = \begin{cases} 1 & \text{if } |\xi| \le \frac{1}{2}, \\ 0 & \text{if } |\xi| > \frac{1}{2}. \end{cases}$$

The operator T_λ, defined in (2.29), has in this case the form

$$T_\lambda f(x) = f_0 = (2\pi)^{-N} \int_{\mathbb{T}^N} f(y)dy$$

and is continuous in $L_p(\mathbb{T}^N)$ for all $p \geq 1$. However $\lambda(\xi)$ is not a multiplier in $L_p(\mathbb{R}^N)$, $p \neq 2$, which follows from Fefferman's theorem on spherical multipliers (cf. Sect. 2.2, Chapter 2). This depends on the fact that the class of multipliers in $L_p(\mathbb{R}^N)$ is dilation (homothety) invariant: if $\varphi \in M_p(\mathbb{R}^N)$ and $\varphi_\varepsilon(\xi) = \varphi(\varepsilon \cdot \xi)$, $\varepsilon > 0$, then $\varphi_\varepsilon \in M_p(\mathbb{R}^N)$. Moreover, the norms of the corresponding operators do not depend on ε. In fact, let

$$T_\varphi f(x) = (2\pi)^{-N/2} \int_{\mathbf{R}^N} \varphi(\xi)\hat{f}(\xi)e^{ix\xi}d\xi,$$

and write $E\varphi(\xi) = \varphi(\varepsilon\xi)$. Then, as is readily seen,

$$T_{E\varphi}f(x) = E^{-1}T_\varphi Ef(x),$$

so that

$$\begin{aligned}\|T_{E\varphi}f\|_{L_p(\mathbf{R}^N)} &= \|E^{-1}T_\varphi Ef\|_{L_p(\mathbf{R}^N)} = \varepsilon^{\frac{N}{p}}\|T_\varphi Ef\|_{L_p(\mathbf{R}^N)} \\ &\leq \varepsilon^{\frac{N}{p}}\|T_\varphi\| \cdot \|Ef\|_{L_p(\mathbf{R}^N)} = \|T_\varphi\| \cdot \|f\|_{L_p(\mathbf{R}^N)},\end{aligned}$$

i. e. $\|T_{E\varphi}\| \leq \|T_\varphi\|$.

In an analogous way one establishes the opposite inequality.

It is clear that sequences $\lambda(n) \in M_p(\mathbb{T}^N)$ in general do not have such properties. Therefore, for the converse of de Leeuw's theorem we have to impose additional requirements, making this circumstance more precise.

Theorem (cf. Stein and Weiss (1971), Chapter VII). *Let $\lambda(\xi)$ be a continuous function on $\mathbb{R}^N$. Let us assume that for each $\varepsilon > 0$ the sequence $\lambda_\varepsilon(n) = \lambda(\varepsilon n)$ is a multiplier on $L_p(\mathbb{T}^N)$, $1 \leq p \leq \infty$, with the norm depending uniformly on ε. Then $\lambda(\xi)$ is a multiplier on $L_p(\mathbb{R}^N)$.*

3.4. Pseudo-Differential Operators. The theory of multipliers is closely connected with the theory of partial differential operators. Let

$$P(D) = \sum_{|\alpha| \leq m} c_\alpha D^\alpha \tag{2.42}$$

be a differential polynomial with constant coefficients. Recall that $D^\alpha = D_1^{\alpha_1} \ldots D_N^{\alpha_N}$, where $D_j = \frac{1}{i}\frac{\partial}{\partial x_j}$.

The imaginary unit has been included in D_j in order to simplify the action of the Fourier integral expansion. With such a normalization we have

$$D^\alpha f(x) = (2\pi)^{-N/2} \int_{\mathbf{R}^N} \xi^\alpha \hat{f}(\xi)e^{ix\xi}d\xi.$$

The action of the operator (2.42) takes the following form

$$P(D)f(x) = (2\pi)^{-N/2} \int_{\mathbf{R}^N} P(\xi)\hat{f}(\xi)e^{ix\xi}d\xi. \tag{2.43}$$

Let us consider the differential equation

$$P(D)f(x) = g(x), \quad x \in \mathbb{R}^N. \tag{2.44}$$

If the functions f and g drop off sufficiently fast at infinity then, passing to Fourier transforms and taking account of (2.43), we obtain

$$P(\xi)\hat{f}(\xi) = \hat{g}(\xi),$$

whence

$$f(x) = (2\pi)^{-N/2} \int_{\mathbf{R}^N} \frac{1}{P(\xi)} \hat{g}(\xi) e^{ix\xi} d\xi. \tag{2.45}$$

From formula (2.45) one can also find the partial derivatives of f:

$$D^\alpha f(x) = (2\pi)^{-N/2} \int_{\mathbf{R}^N} \frac{\xi^\alpha}{P(\xi)} \hat{g}(\xi) e^{ix\xi} d\xi. \tag{2.46}$$

If $\varphi(\xi) = \frac{\xi^\alpha}{P(\xi)}$ is a multiplier on $L_p(\mathbb{R}^N)$ then we get from (2.46) the estimate

$$\|D^\alpha f\|_{L_p(\mathbf{R}^N)} \leq \text{const} \cdot \|g\|_{L_p(\mathbf{R}^N)}. \tag{2.47}$$

Estimates of this type play a great rôle in the study of equation (2.44). We will not enter into questions connected with the strict foundation of the above reasoning, in particular, the problem of division with the algebraic polynomial $P(\xi)$. It is clear that the best estimates of the type (2.47) can be obtained for polynomials having in some sense a small number of real zeros, i. e. polynomials for which the set

$$N(P) = \{\xi \in \mathbb{R}^N : P(\xi) = 0\}$$

is not too big. For example, for a homogeneous elliptic polynomial $P(\xi)$, i. e. a polynomial satisfying the condition

$$P(\xi) \geq c_0 |\xi|^m,$$

$N(P)$ consists of the origin only. For such a polynomial, $\frac{\xi^\alpha}{P(\xi)}$ is a multiplier in L_p for all α, $|\alpha| \leq m$, so that by (2.46)

$$\|f\|_{W_p^m(\mathbf{R}^N)} \leq \text{const} \cdot \|g\|_{L_p(\mathbf{R}^N)}. \tag{2.48}$$

This estimate is best possible, as the exponent m in the left hand side can not be replaced by $m + \varepsilon$. Moreover, if (2.48) is fulfilled for all solutions of (2.44) (i. e. solutions corresponding to an arbitrary $g \in L_p(\mathbb{R}^N)$), then the operator P must be elliptic.

The ease with which the Fourier transforms converts differential operations into algebraic operations leads to the wish to work out a similar method for solutions of equations with variable coefficients. Let us try to apply the same reasoning to the operator

$$P(x, D) = \sum_{|\alpha| \leq m} c_\alpha(x) D^\alpha.$$

Proceeding as above, we obtain

$$P(x, D) f(x) = (2\pi)^{-N/2} \sum_{|\alpha| \leq m} c_\alpha(x) \int_{\mathbb{R}^N} \xi^\alpha \hat{f}(\xi) e^{ix\xi} d\xi =$$

$$= (2\pi)^{-N/2} \int_{\mathbb{R}^N} P(x, \xi) \hat{f}(\xi) e^{ix\xi} d\xi.$$

The factor in front of the exponential in the last integral is no longer the Fourier transform of the function to the left. However, in this case one can introduce the operator

$$Qf(x) = (2\pi)^{-N/2} \int_{\mathbb{R}^N} \frac{1}{P(x, \xi)} \hat{f}(\xi) e^{ix\xi} d\xi$$

and show that it is an "almost inverse" of the operator $P(x, \xi)$.

Viewing the formula

$$P(x, D) f(x) = (2\pi)^{-N/2} \int_{\mathbb{R}^N} P(x, \xi) \hat{f}(\xi) e^{ix\xi} d\xi \tag{2.49}$$

as the *definition* of the operator $P(x, \xi)$ leads to the idea of extending the definition to functions $P(x, \xi)$ of a more general type. This route leads to the notion of *pseudo-differential operator.*

Let us define the class of symbols S^m as the set of functions $p(x, \xi) \in C^\infty(\mathbb{R}^N \times \mathbb{R}^N)$ such that for each compact set $K \subset \mathbb{R}^N$ and all multi-indices α, β one has the inequality

$$|D_x^\alpha D_\xi^\beta p(x, \xi)| \leq \text{const} \cdot (1 + |\xi|)^{m - |\beta|} \tag{2.50}$$

with a constant which does not depend on $x \in K$ and $\xi \in \mathbb{R}^N$.

Formula (2.49) assigns to each symbol $p(x, \xi) \in S^m$ a pseudo-differential operator $p(x, D)$. Their merit is that the action of such operators actually leads to algebraic operators on their symbols. For example, if $P(x, D)$ is an *elliptic pseudo-differential operator*, i. e.

$$P(x, \xi) \geq c_0 |\xi|^m$$

for sufficiently large $|\xi|$, then the inverse operator likewise is a pseudo-differential operator and its symbol is "principally" equivalent to $\frac{1}{P(x,\xi)}$.

Let us point out that condition (2.50) with respect to ξ, reminds of condition (2.33) in the nonperiodic analogue of Marcinkiewicz's theorem. Of course, this is not coincidental, because the symbol class $S^0(\mathbb{R}^N)$ has multiplier properties. Indeed, if $P(x,\xi) \in S^0(\mathbb{R}^N)$, then for any function $f \in L_p(\mathbb{R}^N)$, $1 < p < \infty$, with compact support one has $P(x,D)f \in L_p(K)$ for any compact set K (cf. Taylor (1981), Chapter XI, Sect. 2). Let us remark that the local character of this result is connected with the conventional form of condition (2.50). If this condition holds uniformly in $x \in \mathbb{R}^N$ (and not only on compact sets) then the corresponding operator $P(x,D)$ is continuous on $L_p(\mathbb{R}^N)$, $1 < p < \infty$. One has analogous theorems for operators whose symbols are subject to less restrictive conditions than (2.50) (cf. Taylor (1981), Chapter XI).

Thus, we have established a connection between zero order pseudo-differential operators and multipliers in $L_p(\mathbb{R}^N)$. A special case of zero order pseudo-differential operators are singular integral operators.

The Hilbert transform (cf. (2.16)) is a *one-dimensional singular operator* and that it is a pseudo-differential operator follows from (2.17). In order to pass to the multivariate situation let us rewrite the operator (2.17) as

$$Hf(x) = \lim_{\varepsilon\to 0} \frac{1}{\pi} \int_{|t|\geq\varepsilon} \frac{\operatorname{sign} t}{|t|} f(x-t)dt.$$

The order of the corresponding kernel equals the dimension of the space, i. e. is equal to 1. Therefore it is essential that the function sign t satisfies the "cancellation condition" $\int_{-A}^{A} \operatorname{sign} t\, dt = 0$.

Now let $\Omega(x)$ be a function in N variables which is homogenous of degree 0, i. e. $\Omega(\varepsilon x) = \Omega(x)$, $\varepsilon > 0$. This is the same as to say that Ω is constant on rays issuing from the origin. In particular, Ω is completely determined by its restriction to the unit sphere S^{N-1}. Assume further that Ω satisfies the cancellation condition

$$\int_{S^{N-1}} \Omega(x)d\sigma = 0$$

and a condition of "Dini type": if

$$\sup_{|x-x'|\leq\delta,x,x'\in S^{n-1}} |\Omega(x) - \Omega(x')| = \omega(\delta),$$

then

$$\int_0^1 \frac{\omega(\delta)}{\delta} d\delta < \infty.$$

Let us define the operator T_ε with the aid of the formula

$$T_\varepsilon f(x) = \int_{|y|\geq\varepsilon} \frac{\Omega(y)}{|y|^N} f(x-y)dy, \quad f \in L_p(\mathbb{R}^N),$$

The following statements are valid (cf. Stein (1970)):

1) the limit

$$Tf(x) = \lim_{\varepsilon \to 0} T_\varepsilon f(x)$$

exists in L_p, $1 < p < \infty$;

2) the operator T is bounded in $L_p(\mathbb{R}^N)$, $1 < p < \infty$;

3) there exists a bounded measurable function (a multiplier) $m(y)$ such that $(\widehat{Tf})(y) = m(y)\hat{f}(y)$ for all $f \in L_2(\mathbb{R}^N)$, where $m(y)$ is of degree 0 and has the form

$$m(y) = \int_{S^{N-1}} \left[\frac{\pi i}{2} \operatorname{sign}\,(xy) + \log \frac{1}{|xy|} \right] \Omega(x) d\sigma(x),\ |y| = 1.$$

The operator T is called a *singular integral operator.*

Next let us consider the limit $T_\varepsilon f$, $\varepsilon \to 0$, in the sense of a. e. convergence. As in other cases when a. e. convergence is studied, it is expedient to invoke the corresponding *maximal operator*:

$$T^* f(x) = \sup_\varepsilon |T_\varepsilon f(x)|.$$

An important property of the operator T^* is that it is of strong type (p, p) for $1 < p < \infty$, i. e. there exists a constant $c_p > 0$ such that (cf. Stein (1970))

$$\|T^* f\|_{L_p(\mathbb{R}^N)} \le c_p \|f\|_{L_p(\mathbb{R}^N)}. \tag{2.51}$$

We remark that in the proof of estimates for the maximal operator a major rôle is played by the *maximal function*

$$Mf(x) = \sup_{r>0} \frac{1}{|B(x,r)|} \int_{|B(x,r)|} |f(y)| dy,$$

(where $B(x, r)$ is a ball of radius r and center x), for which the following estimate is fulfilled:

$$\|Mf\|_{L_p(\mathbb{R}^N)} \le c_p \|f\|_{L_p(\mathbb{R}^N)}, \quad 1 < p \le \infty$$

(the maximal function and its properties are studied in detail, e. g., in Stein's monograph (1970)).

Using the estimate (2.51) the existence of $\lim_{\varepsilon \to 0} T_\varepsilon f(x)$ a. e. can be proved rather easily.

If $f \in C_0^\infty(\mathbb{R}^N)$ then $T_\varepsilon f(x)$ converges uniformly on $\mathbb{R}^N$ as $\varepsilon \to 0$. Indeed

$$T_\varepsilon f(x) = \int_{|y| \ge 1} \frac{\Omega(y)}{|y|^N} f(x-y) dy + \int_{\varepsilon \le |y| \le 1} \frac{\Omega(y)}{|y|^N} [f(x-y) - f(x)] dy$$

(by the cancellation condition). The first integral is a continuous function on $\mathbb{R}^N$ while the second converges uniformly in x as $\varepsilon \to 0$ in view of the differentiability of f.

Denoting by $\Lambda f(x)$ the oscillation of the function $T_\varepsilon f$ as $\varepsilon \to 0$:

$$\Lambda f(x) = \overline{\lim_{\varepsilon\to 0}} T_\varepsilon f(x) - \underline{\lim_{\varepsilon\to 0}} T_\varepsilon f(x),$$

it is clear that $\Lambda f(x) \leq 2T^* f(x)$. Let us write $f = f_1 + f_2$, where $f_1 \in C_0^\infty(\mathbb{R}^N)$ and $\|f_2\|_{L_p(\mathbf{R}^N)} \leq \delta$. It follows from (2.51) that $\|\Lambda f\|_{L_p} \leq c p\delta$ so that $\Lambda f(x) = 0$ a. e. and, consequently, $\lim_{\varepsilon\to 0} T_\varepsilon f(x)$ exists a. e. provided $1 < p < \infty$. That the limit is precisely $Tf(x)$ follows from the strong convergence $T_\varepsilon \to T$ in $L_p(\mathbb{R}^N)$.

It will be expedient to make the following remarks concerning a. e. convergence. The preceding scheme is very general and is often encountered. The proof of a. e. convergence breaks up into two parts (this refers also to the results set forth in the following chapter of this part). One of them is very deep and includes the essence of the result. This part is expressed in terms of inequalities for the maximal operators. The second part is as a rule somewhat simpler: it amounts to proving the a. e. convergence on a dense subset of the space under consideration.

Singular integral operators have numerous applications in the theory of multiple Fourier series and integrals, and in mathematical physics (for example, in the study of boundary values of harmonic functions and in the study of generalized Cauchy-Riemann equations; for details see the monographs Garcia-Cuerva and Rubio de Francia (1972), de Guzman (1981), Hörmander (1983–85), Taylor (1981), Stein and Weiss (1971)). Moreover, it was precisely the study of singular integral operators which paved the way for the introduction of the notion of pseudo-differential operator (cf., for instance, Hörmander (1983–85)).

3.5. Fourier Integral Operators. One is led to operators of a more general type if one looks in a different way at the definition (2.49). Writing out in the integral the value of $\hat{f}(\xi)$ we obtain

$$P(x,D)f(x) = (2\pi)^{-N} \int_{\mathbf{R}^N} \int_{\mathbf{R}^N} P(x,\xi) e^{i(x-y)\xi} f(y)\,dy\,d\xi.$$

If we replace the symbol $P(x,\xi)$ by the amplitude $a(x,y;\xi)$, we obtain as a result

$$Af(x) = \int_{\mathbf{R}^N} \int_{\mathbf{R}^N} a(x,y;\xi) e^{i(x-y)\xi} f(y)\,dy\,d\xi.$$

At first sight the class of such operators looks much bigger than the class of pseudo-differential operators but it turns out that it is not so. Indeed (cf. Taylor (1981), Chapter II, Sect. 3), for any amplitude $a(x,y;\xi)$ one can find a symbol $P(x,\xi)$ such that

$$A = P(x,D).$$

Operators which are definitely more general than pseudo-differential operators can be obtained if one replaces $(x-y)\xi$ in the exponential by a phase function $\psi(x,y,\xi)$:

$$Af(x) = \int_{\mathbb{R}^N}\int_{\mathbb{R}^N} a(x,y;\xi)e^{i\psi(x,y,\xi)}f(y)dyd\xi. \tag{2.52}$$

The phase function is assumed to be real, smooth and homogeneous of degree one in ξ. The operators (2.52), which were called *Fourier integral operators* by Hörmander, were introduced by him in (1968) in the study of eigenfunction expansions of elliptic operators with variable coefficients. Operators of a more general form than (2.52) were constructed by V. P. Maslov (1965) in connection with the development of asymptotic methods in the study of problems of mathematical physics.

Since the paper of Hörmander (1968) one has conjectured that the operator of partial sums of the eigenfunction expansion of an elliptic operator $p(x,D)$ equals asymptotically

$$E_\lambda f(x) \sim (2\pi)^{-N} \int_{p(x,\xi)<\lambda}\int_{\mathbb{R}^N} e^{i(x-y)\xi}f(y)dyd\xi,$$

but one has not been able to obtain satisfactory estimates for the remainder. Hörmander proved that the principal part of the operator E_λ can be represented by the Fourier integral operator

$$E_\lambda f(x) \sim (2\pi)^{-N} \int_{p(x,\xi)<\lambda}\int_{\mathbb{R}^N} e^{\psi(x,y,\xi)}f(y)dyd\xi,$$

with a phase function ψ defined with the aid of the symbol of the operator $P(x,D)$.

The study of Fourier integral operators of the type (2.52) constitutes a problem which is considerably harder than the corresponding problem in the case of pseudo-differential operators. One of the main difficulties is connected with the fact that the phase function ψ is not uniquely determined by A. The function $\psi(x,y,\xi)$ can be defined on $X\times Y\times\mathbb{R}^N$ where X, Y and $\mathbb{R}^N$ need not necessarily coincide; in particular, they might all have different dimension. Furthermore, the amplitude can in different representations of A be defined on different sets, which complicates the problem of defining the "principal symbol" of A.

Closing this chapter devoted to the L_p-theory, we remark that the problem of the continuity of Fourier integral operators in L_p is still far from being solved.

Chapter 3
Convergence Almost Everywhere

§1. Rectangular Convergence

1.1. Quadratic Convergence. In 1915 in his dissertation "Integrals and trigonometric series" N. N. Luzin made the conjecture that the Fourier series of any function in $L_2(\mathbb{T}^1)$ converges a. e. In other words, if

$$\sum |c_n|^2 < \infty, \tag{3.1}$$

then the series

$$\sum c_n e^{inx} \tag{3.2}$$

converges for almost every $x \in \mathbb{T}^1$.

Over a period of many years the *Luzin conjecture* attracted the attention of the specialists inducing a large number of investigations. This long story was closed only in 1966 when Carleson (1966) published a remarkable result. He proved that the Fourier series of any function in $L_2(\mathbb{T}^1)$ converges a. e., that is, Luzin's problem has a positive solution.

It is easy to extend Carleson's theorem to N-fold Fourier series with quadratic or cubic summation (cf. Tevzadze (1970)). Let us indicate what this is about if $N = 2$. Let $f \in L_2(\mathbb{T}^2)$ be expanded in a double Fourier series whose quadratic partial sums equal

$$S_k(x_1, x_2) = \sum_{|n_2|\le k} \sum_{|n_1|\le k} f_{n_1 n_2} e^{i(n_1 x_1 + n_2 x_2)}.$$

Let us divide this sum into "one dimensional" partial sums with coefficients such that the series of their squares is convergent. To this end we divide the "square"

$$Q_k = \{(n_1, n_2) \in \mathbb{Z}^2 : |n_1| \le k, |n_2| \le k\}$$

into "triangles" with a common vertex at the origin and set

$$A_{n_1}(x_2) = \sum_{|n_2|\le|n_1|} f_{n_1 n_2} e^{in_2 x_2}, \quad B_{n_2}(x_1) = \sum_{|n_1|<|n_2|} f_{n_1 n_2} e^{in_1 x_1}.$$

Then clearly

$$S_k(x_1, x_2) = \sum_{|n_1|\le k} A_{n_1}(x_2) e^{in_1 x_1} + \sum_{|n_2|\le k} B_{n_2}(x_1) e^{in_2 x_2}. \tag{3.3}$$

As

$$\sum_{n_1=-\infty}^{\infty} \int_{-\pi}^{\pi} A_{n_1}^2(x_2)\,dx_2 = \sum_{n_1=-\infty}^{\infty} \sum_{|n_2|\le|n_1|} |f_{n_1 n_2}|^2 < \infty,$$

then by Levi's theorem the series

$$\sum_{n_1=-\infty}^{\infty} A_{n_1}^2(x_2)$$

will converge for almost every $x_2 \in \mathbb{T}^1$. Therefore, the first of the sums in the right hand side of (3.3) is for almost every $x_2 \in \mathbb{T}^1$ the partial sum of a Fourier series of a function in $L_2(\mathbb{T}^1)$ and, by Carleson's theorem, it must then converge for almost all $x_1 \in \mathbb{T}^1$. In an analogous way one shows that the second sum in (3.3) for almost every $x_1 \in \mathbb{T}^1$ converges in x_2 a. e. on $\mathbb{T}^1$, i. e.

$$\lim_{k\to\infty} S_k(x_1, x_2) = f(x_1, x_2)$$

a. e. on $\mathbb{T}^2$. (That the series converges to f follows from the fact that S_k converges to f in L_2 and therefore in measure, so that a subsequences $S_{k(j)}$ must converge to f a. e. on $\mathbb{T}^1$.)

If $N > 2$ the proof is similar. Then the "cube"

$$Q_k = \{n \in \mathbb{Z}^N : |n_j| \leq k, j = 1, 2, \ldots, N\}$$

has to be divided into "pyramides" with a common vertex at the origin. Indeed, if $f \in L_2(\mathbb{T}^N)$ has the multiple Fourier expansion

$$f(x) \sim \sum_{n\in\mathbb{Z}^N} f_n e^{inx},$$

then the cubic partial sum can be written as

$$S_k(x) = \sum_{l=1}^{N} \sum_{|n_l|\leq k} A_{n_l}(x^l) e^{in_l x_l}. \tag{3.4}$$

Here we have introduced for $x = (x_1, x_2, \ldots, x_N) \in \mathbb{R}^N$ the notation $x^l = (x_1, x_2, \ldots, x_{l-1}, x_{l+1}, \ldots, x_N) \in \mathbb{R}^{N-1}$ putting

$$A_{n_l}(x^l) = \sum_{|n^l|\leq|n_l|} f_n e^{in^l x^l};$$

note, the inequality for the indices must in some cases be taken to be strict. In other respects the proof does not differ from the proof in the two dimensional case.

Let us point out that the coefficients $A_{n_l}(x^l)$ in the inner sum (3.4) are not allowed to depend on k, because otherwise they will not be partial sums of a Fourier series and their behavior will not be governed by Carleson's theorem. It is precisely this circumstance that dictates the way of writing the cubical sum in the form (3.4).

Carleson's result has been improved by Hunt (1968), who showed that the Fourier series of any function $f \in L_p(\mathbb{T}^1)$, $p > 1$, converges a. e. on $\mathbb{T}^1$. On the other hand, as early as in 1922 A. N. Kolmogorov constructed a remarkable example of a function whose Fourier series diverges a. e. and even at every point. Thus, in the classes $L_p(\mathbb{T}^1)$ the problem of a. e. convergence is completely solved.

Hunt's theorem has also been extended to multiple Fourier series with quadratic summation. There is even a somewhat stronger result due to Sjölin (1971). He showed that if

$$\int_{\mathbb{T}^N} |f(x)| \left(\log^+ |f(x)|\right)^N \log^+ \log^+ |f(x)| dx < \infty, \tag{3.5}$$

then the quadratic partial sums of the Fourier series of f converge to f a. e. on $\mathbb{T}^N$. Here $\log^+ t = \log(\max\{1, t\})$. We remark that this theorem is new also for $N = 1$. As $\log t$ grows slower than any power of t as $t \to \infty$, every function in $L_p(\mathbb{T}^N)$, $p > 1$, satisfies condition (3.5). Therefore it follows from Sjölin's theorem that the problem of a. e. convergence of multiple Fourier series with quadratic summability in the classes $L_p(\mathbb{T}^N)$ is completely solved (for $p = 1$ a counterexample is provided by the function constructed by Kolmogorov).

1.2. Rectangular Convergence. The problem of quadratic a. e. convergence of functions in L_p does not depend on the number of dimensions and the result has exactly the same formulation for $N > 1$ as in the one dimensional case. However, rectangular convergence leads to completely new phenomena.

In (1970) Fefferman gave an example of a continuous periodic function $f(x_1, x_2)$ whose Fourier series when summed rectangularly diverges unboundedly at each interior point of the square $\mathbb{T}^2$. In this example some of the paradoxical properties of rectangular partial sums, casually mentioned in the introduction, become manifest.

Let us illustrate this at the hand of the following remark. Let

$$S_{m_1 m_2}(x_2, x_2) = \sum_{|n_1| \le m_1} \sum_{|n_2| \le m_2} f_{n_1 n_2} e^{i(n_1 x_1 + n_2 x_2)}$$

be the partial sums of the Fourier series of the function $f \in L_2(\mathbb{T}^2)$ and let $m_1(k)$, $m_2(k)$ be arbitrary decreasing sequences of integers. Then for $k \to \infty$

$$S_{m_1(k) m_2(k)}(x_1, x_2) \to f(x_1, x_2)$$

a. e. on $\mathbb{T}^2$. The proof of this statement is done in complete analogy to the reasonings in Subsect. 1.1 of this section. Consequently, each sequence of rectangular partial sums, taken with respect to any fixed exhausting system of rectangles, converges a. e. As $C(\mathbb{T}^2) \subset L_2(\mathbb{T}^2)$ this remark applies also to the Fourier series of the function constructed by Fefferman. Nevertheless, the rectangular partial sums of its Fourier series diverge at each point.

There is of course no contradiction here. What then is the point? It is a question on the sets of measure zero where the sequence $S_{m_1(k)m_2(k)}(x_1, x_2)$ diverges. If there were no such sets of measure zero, i. e. if the convergence occured at all points, then we would indeed get a contradiction, because a numerical multiple series with rectangular summation cannot both converge and diverge. In Fefferman's example one can find at each point $(\bar{x}_1, \bar{x}_2) \in \mathbb{T}^2$ a suitable subsequence $(m_1(k), m_2(k))$ such that the sequence $S_{m_1(k)m_2(k)}(x_1, x_2)$ diverges precisely at the point $(\bar{x}_1, \bar{x}_2)$ (but converges at all other points). In other words, to each subsequence $(m_1(k), m_2(k))$ there corresponds a divergence set of measure zero, and these sets taken together cover the whole square $\mathbb{T}^2$.

Thus, for all $p \geq 1$ the problem of a. e. convergence of rectangular sums of Fourier series of functions in $L_p(\mathbb{T}^N)$ has a negative answer for $N > 1$. In particular, Luzin's conjecture concerning the a. e. convergence of the series (3.2) (in the hypothesis of (3.1)) is not true for N-fold series under rectangular summation.

The problem of a. e. convergence is often (especially in the theory of orthogonal series) formulated in terms of Weyl multipliers. A monotone sequence $\lambda(n)$, $n \in \mathbf{Z}^N$, is called a *Weyl multiplier* for a. e. convergence of Fourier series if the condition

$$\sum_{n\in\mathbf{Z}^N} |f_n|^2 \lambda(n) < \infty \tag{3.6}$$

entails the a. e. convergence of the Fourier series of $f \in L_2(\mathbb{T}^N)$. Then *Luzin's problem* can be formulated as follows: is it true that the sequence $\lambda(n) \equiv 1$ is a Weyl multiplier for a. e. convergence? For quadratic summation the answer is, as we have remarked, positive, while for rectangular convergence it is negative.

Which are the exact Weyl multipliers for rectangular summation? So far the answer to this question is known for $N = 2$.

Theorem 1 (Sjölin (1971)). *If*

$$\lambda(n) = \log^2(\min\{|n_1|^2 + 2, \quad |n_2|^2 + 2\}),$$

then (3.6) *entails the a. e. convergence on* $\mathbb{T}^2$ *of the rectangular partial sums of the Fourier series of* f.

Theorem 2 (E. M. Nikishin (1972a)). *Let there be given a sequence* $\lambda(n)$ *such that the assumption* (3.6) *always implies the a. e. convergence on* $\mathbb{T}^2$ *of the rectangular partial sums of the Fourier series of* f. *Then there exist a constant* $c > 0$ *such that*

$$\lambda(n) > c \cdot \log^2(\min\{|n_1|^2 + 2, \quad |n_2|^2 + 2\}).$$

To the problem of a. e. convergence pertains the question of *generalized localization*, as studied by I. L. Bloshanskiĭ. Recall that the validity of the principle of localization means that the convergence of the Fourier series of the function f depends only on the behavior of f in a small neighborhood of the point. More exactly, if $f = 0$ in an open set $\Omega \subset \mathbb{T}^N$ then the Fourier series of f converges to 0 at every point of Ω. For generalized localization one requires that the convergence to 0 occurs a. e. on Ω.

Up to this moment, all what we have said might give the impression that the main problem of the theory of multiple series is the extension of "one-dimensional" results to the N-dimensional case and that the results are classified according to the following test: do they extend to N-fold series or not. When passing from $N = 2$ to $N \geq 3$ such a dichotomy does not take place.

Generalized localization of rectangular partial sums belongs indeed to those properties which discriminate between two- and three-dimensional Fourier series. In Bloshanskiĭ (1975) it is proved that generalized localization holds in the classes $L_p(\mathbb{T}^N)$, $p > 1$, for $N = 2$ but not for $N \geq 3$. In Chapter 1 we have noted an heuristic idea concerning to the behavior of rectangular partial sums: if under the hypothesis of certain conditions convergence does not hold in $\mathbb{T}^{N-1}$ then under the same assumptions localization does not take place in $\mathbb{T}^N$. In some sense even more is true: if convergence takes place in $\mathbb{T}^{N-1}$ then one has also localization in $\mathbb{T}^N$. These intuitive conclusions can be confirmed. Rectangular partial sums of Fourier series of functions in the classes $L_p(\mathbb{T}^N)$, $p > 1$, converge if $N = 1$ (the result of Carleson and Hunt) and diverge for $N = 2$ (Fefferman's example). Let us indicate how one can obtain the results of Bloshanskiĭ (1975) from this.

If the function $f(x_1, x_2) \in L_p(\mathbb{T}^2)$ vanishes for $|x_1| < \delta$, $|x_2| < \delta$, then it can be written as the sum of two functions, one of which vanishes for $|x_1| < \delta$, the other for $|x_2| < \delta$. For a function of this type the rectangular partial sum can be written as follws:

$$S_{m_1 m_2}(x_1, x_2) = \pi^{-2} \int_{\delta \leq |s| \leq \pi} D_{m_1}(s) \int_{-\pi}^{\pi} D_{m_2}(t) f(x_1 + s, x_2 + t)\, dt\, ds.$$

In view of Hunt's theorem (Sect. 1.1), the inner integral converges for a. e. points $(x_1 + s, x_2) \in \mathbb{T}^2$ and $m_2 \to \infty$ to $f(x_1 + s, x_2)$; here we must take into account that in view of Fubini's theorem the function $f(x_1, x_2)$ belongs for a. e. $x_1 \in \mathbb{T}^1$ to $L_p(\mathbb{T}^1)$ in x_2. In the outer integral the kernel $D_{m_1}(s)$ is uniformly bounded in the domain of integration so that it is natural to expect that the outer integral too converges for $m_2 \to \infty$. After some not very hard technical calculations the above reasoning can be given a solid foundation.

The absence of generalized localization in $L_p(\mathbb{T}^3)$ (and also in $C(\mathbb{T}^3)$) can be established with the aid of Fefferman's example of a continuous function $f_0(x_1, x_2)$ whose Fourier series summed rectangularly diverges at each point $(x_1, x_2) \in \mathbb{T}^2$. As in Subsect. 1.2 of Sect. 1, Chapter 1, let us set

$$f(x) = f_0(x_1, x_2) \cdot h(x_3), \quad x = (x_1, x_2, x_3),$$

where $h \in C(\mathbb{T}^1)$ is chosen such that $h(t) = 0$ for $|t| \leq \delta$, while the partial sums $S_k(t, h)$ of its Fourier series at each point $t \in \mathbb{T}^1$ are different from zero for an infinite number of indices. The function f vanishes in the strip $|x_3| < \delta$ but its rectangular partial sums

$$S_{m_1m_2m_3}(x, f) = S_{m_1m_2}(x_1, x_2, f_0) \cdot S_{m_3}(x_3, h), \quad x \in \mathbb{T}^1,$$

diverge everywhere on this strip (and even everywhere on $\mathbb{T}^3$). This is not hard to realize if we take for each m_3 the indices m_1, m_2 large enough. Let us also remark that the problem of generalized localization of quadratic or cubic partial sums, which also has been studied in detail by I. L. Bloshanskiĭ (1976), is much harder.

We have spoken about the classes $L_p(\mathbb{T}^N)$ for $p > 1$. If $p = 1$, generalized localization for rectangular partial sums does not hold for any $N \geq 2$, which readily follows from Kolmogorov's example of a function in $L_1(\mathbb{T}^1)$ with an unboundedly divergent Fourier series.

§2. Convergence a. e. of Spherical Sums and Their Means

2.1. Convergence of Spherical Sums. Recall that the spherical sums of a function f take the form

$$E_\mu f(x) = (2\pi)^{-N} \sum_{|n| \leq \mu} f_n e^{inx}. \tag{3.7}$$

One of the first questions which arise in the study of a. e. convergence of the sums (3.7) is the question of the validity of the Luzin conjecture: is it true that the spherical sums (3.7) of the Fourier series of an arbitrary function $f \in L_2(\mathbb{T}^N)$ converge a. e. on $\mathbb{T}^N$? In other words, does Carleson's theorem extend to N-fold Fourier series when the latter is summed spherically? The answer to this question is open so far.

What is known is only that Hunt's theorem (Subsect. 1.1 of Sect. 1) does not extend to N-fold ($N > 1$) series summed by circles. This follows from general results by B. S. Mityagin and E. M. Nikishin (cf. (1973a), (1973b)) on the divergence of spectral expansions and lower estimates of partial sums. Indeed, for each $p \in [1, 2]$ there exists a function $f \in L_p(\mathbb{T}^N)$ such that on a set of postive measure

$$\overline{\lim_{\mu \to \infty}} |E_\mu f(x)| = \infty.$$

E. M. Nikishin has remarked that this result (slightly less general than in (1973a), (1973b)) is an immediate consequence of Fefferman's theorem to the effect that the multiple trigonometric system is not a basis in $L_p(\mathbb{T}^N)$, $1 \leq p < 2$, under spherical summation. To explain this we resort to Stein's theorem on sequences of operators.

Let us consider a sequence of translation invariant linear operators $T_k : L_p(\mathbb{T}^N) \to L_p(\mathbb{T}^N)$, $k = 1, 2, \ldots$, and let us introduce the *maximal operator*

$$T_* f(x) = \sup_k |T_k f(x)|.$$

Theorem (Stein (1970)). *If for each $f \in L_p(\mathbb{T}^N)$, where $1 \le p \le 2$, the function $T_* f(x)$ is finite a. e. on $\mathbb{T}^N$, then T_* is of weak type (p,p).*

Let us assume that for some p, $1 \le p \le 2$, the multiple Fourier series of any function $f \in L_p(\mathbb{T}^N)$ converges a. e. on $\mathbb{T}^N$ under spherical summation. Then for a. e. x (3.7) is bounded in μ and so the maximal operator

$$E_* f(x) = \sup_\mu |E_\mu f(x)| \tag{3.8}$$

is finite a. e. on $\mathbb{T}^N$ for each function $f \in L_p(\mathbb{T}^N)$. Therefore by Stein's theorem (infra) E_* is of weak type (p,p). As $L_2(\mathbb{T}^N) \subset L_p(\mathbb{T}^N)$ for $1 \le p \le 2$ it follows that $E_* f(x)$ is finite a. e. on $\mathbb{T}^N$ also for any $f \in L_2(\mathbb{T}^N)$ and therefore, again by Stein's theorem, E_* must be of weak type (2,2).

Now we can apply the Marcinkiewicz interpolation theorem (cf. Stein (1970), Sect. 2, Chapter 5) according to which E_* is of strong type (p_1, p_1) for every p_1 in the interval $p < p_1 < 2$, i. e.

$$\|E_* f\|_{L_{p_1}(\mathbb{T}^N)} \le \text{const} \cdot \|f\|_{L_{p_1}(\mathbb{T}^N)} .$$

It follows from this inequality and from (3.8) that the operators E_μ are uniformly bounded in $L_p(\mathbb{T}^N)$, contradicting Fefferman's theorem, so that the statement to the effect that the partial sums (3.7) converge a. e. for every $f \in L_p(\mathbb{T}^N)$, $1 \le p < 2$ is not true.

2.2. Convergence a. e. of Spherical Riesz Means. The results on the divergence of the spherical sums (3.7) set forth in the last section force us to turn once more to Riesz means of these sums, which have the form

$$E_\mu^s f(x) = \sum_{|n| \le \mu} \left(1 - \frac{|n|^s}{\mu^2}\right)^s f_n e^{inx}. \tag{3.9}$$

In the study of questions of a. e. convergence it is convenient to introduce the maximal operator

$$E_*^s f(x) = \sup_\mu |E_\mu^s f(x)|. \tag{3.10}$$

As was proved in Sect. 3.4 of Chapter 2 it follows from the boundedness of the maximal operator in $L_p(\mathbb{T}^N)$ that the means (3.9) converge a. e. for every $f \in L_p(\mathbb{T}^N)$.

Let us start with the simplest case $p = 2$. The estimate for the maximal operator in this case follows from the following lemma, which is known as *Kaczmarz's lemma* (cf. Kaczmarz and Steinhaus (1951)).

Lemma. *Assume that* $\operatorname{Re} s > 0$, $\operatorname{Re} \alpha > 0$ *and let F be an arbitrary measurable subset of $\mathbb{T}^N$. Then for every function $f \in L_2(\mathbb{T}^N)$ we have the inequality*

$$\|E_*^s f\|_{L_2(F)} \le \text{const} \left[\|E_*^\alpha f\|_{L_2(F)} + \|f\|_{L_2(\mathbb{T}^N)}\right] . \tag{3.11}$$

Proof. The Riesz means (3.9) of different orders are related by the following formula

$$\lambda^s E_\lambda^s f = \frac{\Gamma(s+1)}{\Gamma(\beta+1)\Gamma(s-\beta+1)} \int_0^\lambda (\lambda-t)^{s-\beta-1} t^\beta E_t^\beta f dt, \tag{3.12}$$

valid for $\operatorname{Re} s > \operatorname{Re}\beta + \frac{1}{2} > 0$. This follows readily using the definition (17).

If we apply the Cauchy-Bunyakovskiĭ inequality to (3.12), we obtain

$$|E_\lambda^s f|^2 \le \text{const} \cdot \frac{1}{\lambda} \int_0^\lambda |E_t^\beta f|^2 dt.$$

From the inequalities

$$\begin{aligned} \frac{1}{\lambda}\int_0^\lambda |E_t^\beta f|^2 dt &\le \frac{2}{\lambda}\int_0^\lambda |E_t^\alpha f|^2 dt + \frac{2}{\lambda}\int_0^\lambda |E_t^\beta f - E_t^\alpha|^2 dt \\ &\le 2|E_*^\alpha f|^2 + 2\int_0^\infty |E_t^\beta f - E_t^\alpha f|^2 \frac{dt}{t} \end{aligned}$$

follows the estimate

$$|E_*^s f|^2 \le c\left\{\int_0^\infty |E_t^\beta f - E_t^\alpha f|^2 \frac{dt}{t} + |E_*^\alpha f|^2\right\}.$$

To finish the proof of (3.11) we have to prove that for $\operatorname{Re}\alpha > -\frac{1}{2}$, $\operatorname{Re}\beta > -\frac{1}{2}$ the following equality holds:

$$\int_0^\infty \|E_t^\beta f - E_t^\alpha f\|_{L_2(\mathbf{T}^N)}^2 \frac{dt}{t} = \|f\|_{L_2(\mathbf{T}^N)}^2 \cdot \int_0^1 |(1-t)^\beta - (1-t)^\alpha|^2 \frac{dt}{t}. \tag{3.13}$$

As

$$\|E_\lambda^\beta f - E_\lambda^\alpha\|_{L_2(\mathbf{T}^N)}^2 = \int_0^\lambda \left|\left(1-\frac{t}{\lambda}\right)^\beta - \left(1-\frac{t}{\lambda}\right)^\alpha\right|^2 d(E_t f, f),$$

we get using Fubini's theorem

$$\begin{aligned} &\int_0^\lambda \|E_\lambda^\beta f - E_\lambda^\alpha\|_{L_2(\mathbf{T}^N)}^2 \frac{d\lambda}{\lambda} \\ &= \int_0^\lambda d(E_t f, f) \cdot \int_t^\infty \left|\left(1-\frac{t}{\lambda}\right)^\beta - \left(1-\frac{t}{\lambda}\right)^\alpha\right|^2 \frac{d\lambda}{\lambda} \\ &= \int_0^\lambda d(E_t f, f) \cdot \int_0^1 [(1-u)^\beta - (1-u)^\alpha]^2 \frac{du}{u}, \end{aligned}$$

concluding the proof of (3.13), and hence also of the lemma.

This proof of Kaczmarz's lemma is due to Peetre (1964).

What is the significance of Kaczmarz's lemma? It means that in the question of a. e. convergence in the class $L_2(\mathbb{T}^N)$ all Riesz means of positive order are equivalent. In other words, if the Riesz means of some order $\alpha > 0$ converge a. e. on $\mathbb{T}^N$, then the Riesz means of any other order $s > 0$ converge also. Moreover, the same proof remains in force for a. e. convergence on an arbitrary measurable subset $F \subset \mathbb{T}^N$. Kaczmarz's lemma holds true not only for multiple Fourier series expansions, but also for arbitrary orthogonal expansions (cf. Alimov, Il'in, and Nikishin (1976/77), Kaczmarz and Steinhaus (1951)).

Thus, it is question of proving the a. e. convergence of Riesz means for some (for example, a very high) order $s > 0$. For $s > \frac{N-1}{2}$ we can use the Poisson summation formula (1.13) or, more exactly, its consequence (1.17), which reduces the problem of the convergence of spherical means of multiple series to the analogous problem for spherical means of Fourier integrals (1.4):

$$R_\mu^s f(x) = \text{const} \cdot \mu^{\frac{N}{2}-s} \int_{\mathbf{R}^N} \frac{J_{\frac{N}{2}+s}(\mu|x-y|)}{|x-y|^{\frac{N}{2}+s}} f(y)\, dy.$$

From this formula and the well-known estimate for the Bessel function

$$|J_\nu(t)| \le \text{const} \cdot \min\{t^{-\frac{1}{2}}, t^\nu\}$$

we obtain after passing to spherical coordinates and integrating by parts

$$|R_\mu^s f(x)| \le \text{const} \cdot \int_0^\infty |H_t f(x)| \cdot (t\mu)^{\frac{N-1}{2}-s} \cdot \min\{1, (t\mu)^{\frac{N+1}{2}+s}\} \frac{dt}{t}. \tag{3.14}$$

Here we have put

$$H_t f(x) = \frac{1}{V_N t^N} \int_{|y|\le t} f(x+y)\, dy, \tag{3.15}$$

where V_N is the volume of the unit ball. Introducing the maximal operator

$$H_* f(x) = \sup_{t>0} |H_t f(x)|, \tag{3.16}$$

we obtain from (3.14) the required estimate

$$R_*^s f(x) \le \text{const} \cdot H_* f(x). \tag{3.17}$$

The maximal operator H_* plays a major rôle in analysis and has been much studied (cf. Stein (1970), Stein and Weiss (1971)). In particular, for any $p > 1$ the operator H_* is of strong type (p,p):

$$\|H_* f\|_{L_p(\mathbf{R}^N)} \le \text{const} \cdot \|f\|_{L_p(\mathbf{R}^N)}. \tag{3.18}$$

It follows from (3.17) and (3.18) that

$$\|R_*^s f\|_{L_p(\mathbb{R}^N)} \leq \text{const} \cdot \|f\|_{L_p(\mathbb{R}^N)}, \quad p > 1, \quad s > \frac{N-1}{2}.$$

This estimate implies that the Riesz means $R_\mu^s f$ with $s > \frac{N-1}{2}$ of any function $f \in L_p(\mathbb{R}^N)$, $p > 1$, converge a. e..

In view of (1.17) the same estimate is true for $E_*^s f$ with $s > \frac{N-1}{2}$:

$$\|E_* f\|_{L_p(\mathbb{T}^N)} \leq \text{const} \cdot \|f\|_{L_p(\mathbb{T}^N)}, \quad p > 1, \quad s > \frac{N-1}{2}, \tag{3.19}$$

while from Kaczmarz's lemma we obtain for $p = 2$ an estimate which is valid for all $s > 0$:

$$\|E_*^s f\|_{L_2(\mathbb{T}^N)} \leq \text{const} \cdot \|f\|_{L_2(\mathbb{T}^N)}, \quad s > 0. \tag{3.20}$$

Consequently, for every $f \in L_2(\mathbb{T}^N)$ the Riesz means $E_\lambda^s f$ of any positive order converge a. e. on $\mathbb{T}^N$. For multiple series this result is due to Mitchell (1951). As we have already told, the question of a. e. convergence of $E_\lambda^s f$, $f \in L_2(\mathbb{T}^N)$, remains open for $s = 0$.

Let us now pass to the case $1 < p < 2$. It follows from (3.19) that for such values of p the Riesz means converge above the critical order. This is a good result for p near 1, but its gets cruder when p approaches 2, as for $p = 2$ a. e. convergence holds true for all $s > 0$. There arises the natural desire to interpolate between (3.19) and (3.20), taking p in (3.19) as close to 1 as possible.

Let us turn attention to the difficulties which arise then. The interpolation theorem of Stein (cf. Sect. 1.3 of Chapter 2), which we would like to apply, pertains to an analytic family of linear operators, but the maximal operator E_*^s is nonlinear. This difficulty is resolved in the following standard way. Denote by M the class of positive measurable functions on $\mathbb{T}^N$ taking finitely many different values. If $\mu \in M$ then by the definition (3.8)

$$|E_{\mu(x)}^s f(x)| \leq E_*^s f(x). \tag{3.21}$$

It is clear that one can pick a sequence $\mu_1 \leq \mu_2 \leq \ldots$ of elements in M such that

$$\lim_{j \to \infty} |E_{\mu_j(x)}^s f(x)| = E_*^s f(x).$$

This allows us to invert (3.21) as follows:

$$\sup_{\mu \in M} \|E_{\mu(x)}^s f(x)\|_{L_p(\mathbb{T}^N)} = \|E_*^s f\|_{L_p(\mathbb{T}^N)}. \tag{3.22}$$

Fix now $\mu \in M$ and consider the linear operator $E^s_{\mu(x)}$, which depends analytically on the parameter s. To this operator it is possible to apply Stein's interpolation theorem, and proceeding as in Sect. 1.3 of Chapter 2, taking acount of (3.21), we obtain from (3.19) and (3.20) the estimate

$$\|E^s_{\mu(x)} f(x)\|_{L_p(\mathbb{T}^N)} \leq \text{const} \cdot \|f\|_{L_p(\mathbb{T}^N)}$$

for all $s > (N-1)(\frac{1}{p} - \frac{1}{2})$, $1 < p \leq 2$, with a constant independent of f, μ. From this estimate and (3.22) follows the desired estimate:

$$\|E^s_* f\|_{L_p(\mathbb{T}^N)} \leq \text{const} \cdot \|f\|_{L_p(\mathbb{T}^N)}, \quad s > (N-1)\left(\frac{1}{p} - \frac{1}{2}\right). \tag{3.23}$$

The estimate (3.23) says that the means $E^s_\lambda f$ of any function $f \in L_p(\mathbb{T}^N)$, $1 < p \leq 2$, converge a. e. for the values of s indicated. This result (as well as the preceding argument) is due to Stein (1958). Let us note that the convergence theorem is valid also for $p = 1$. In this case, the maximal operator H_* is not of strong type (1,1), so the estimate (3.18) is not true, but it is of weak type (1,1), which in view of (3.17) is sufficient for the a. e. convergence of the means $R^s_\mu f$ (and therefore, in view of (1.17), also of the means $E^s_\mu f$) for $s > \frac{N-1}{2}$ for any integrable function f.

To sum up, we may conclude (cf. Stein (1958)) that a sufficient condition for the a. e. convergence of the means $E^s_\mu f$ of a function $f \in L_p(\mathbb{T}^N)$ is that we have

$$s > (N-1)\left(\frac{1}{p} - \frac{1}{2}\right), \quad 1 \leq p \leq 2. \tag{3.24}$$

How sharp is this condition? To elucidate this question we begin with the case of Riesz means of order $s = \frac{N-1}{2}$. Condition (3.24) shows that for the a. e. convergence of Riesz means one requires an order above the critical index only if $p = 1$. That this requirement is essential follows from the following result of Stein's (cf. Stein and Weiss (1971)): there exists a function $f \in L_1(\mathbb{T}^N)$, $n \geq 2$, such that a. e. on $\mathbb{T}^N$

$$\overline{\lim_{\mu \to \infty}} |E^s_\mu f(x)| = +\infty, \quad s = \frac{N-1}{2}.$$

Thus, for $p = 1$ condition (3.24) is sharp. For $p > 1$ one has a result due to K. I. Babenko (1973b): if

$$0 \leq s \leq N\left(\frac{1}{p} - \frac{1}{2}\right) - \frac{1}{2}, \quad 1 \leq p \leq \frac{2N}{N+1}, \tag{3.25}$$

then there exists a function $f \in L_p(\mathbb{T}^N)$ such that $E^s_\mu f$ is divergent on a set of positive measure. We turn the reader's attention to the fact that between the conditions (3.24) and (3.25) there is a gap, which as in the case of L_p convergence is, of course, not accidental (cf. Subsect. 2.1 of this section).

Chapter 4
Fourier Coefficients

§1. The Cantor-Lebesgue Theorem

1.1. The One-Dimensional Case. As is well-known, the convergence of a series implies that its terms must converge to zero. In particular, the convergence of a trigonometric series

$$\sum_{n=1}^{\infty}(a_n \cos nx + b_n \sin nx)$$

at a point $x \in \mathbb{T}^1$ forces the expression $a_n \cos nx + b_n \sin nx$ to tend to zero at this point as $n \to \infty$. But does this imply that $a_n \to 0$, $b_n \to 0$? If it is a question of just one point the answer is, of course, negative, but if the set of such points is sufficiently large the answer may be positive. In terms of measure theory such an answer is provided by the well-known *Cantor-Lebesgue* theorem: if

$$a_n \cos nx + b_n \sin nx \to 0, \quad n \to \infty$$

on a set of positive measure, then $a_n \to 0$, $b_n \to 0$. This theorem, which is rather easy to prove, plays a major rôle in the theory of trigonometric series and, notably, in the problem of the unique representability of functions by convergent trigonometric series.

Before looking for an analogue of the Cantor-Lebesgue theorem for multiple series (which likewise is of great importance in the uniqueness problem) we restate it in the case when the trigonometric series is written in complex form:

$$\sum_{n=-\infty}^{\infty} c_n e^{inx}.$$

In this case it follows from the condition

$$\sum_{|n|=k} c_n e^{inx} \to 0, \quad k \to \infty$$

on a set $E \subset \mathbb{T}^1$ of positive measure that $c_n \to 0$. In this formulation a possible generalization of the Cantor-Lebesgue theorem to N-dimensional series with $N > 1$ becomes quite transparent.

1.2. Spherical Partial Sums. In 1971 Cooke (1971) proved that if $N = 2$ and

$$A_k(x) = \sum_{|n|^2=k} c_n e^{inx} \to 0, \quad k \to \infty \tag{4.1}$$

a. e. on $\mathbb{T}^2$ then

$$\sum_{|n|^2=k} |c_n|^2 \to 0, \quad k \to \infty. \tag{4.2}$$

Cooke's proof is distinguished by its shortness and ingenuity, of which the reader can easily convince himself. The main step consists of an estimate of the norm of A_k in $L_4(\mathbb{T}^2)$ by the norm of the same sum A_k in $L_2(\mathbb{T}^2)$.

It is clear that

$$|A_k(x)|^2 = \sum_{|n|^2=k} \sum_{|m|^2=k} c_n \bar{c}_m e^{i(n-m)x} = \sum_l e^{ilx} \sum_{\substack{n-m=l \\ |n|^2=|m|^2=k}} c_n \bar{c}_m,$$

where the outer sum extends over all vectors l connecting points in the lattice of integers, situated on a sphere of radius $\sqrt{k}$. Let us now apply Parseval's formula:

$$\int_{\mathbb{T}^2} |A_k(x)|^4 dx = 4\pi^2 \left(\sum_{|n|^2=k} |c_n|^2 \right)^2 + 4\pi^2 \sum_{l \neq 0} \left| \sum_{\substack{n-m=l \\ |n|^2=|m|^2=k}} c_n \bar{c}_m \right|^2. \tag{4.3}$$

Until now we have not used the fact that $N = 2$. In particular, an identity analogous to (4.3) holds also for $N > 2$. However, it is only in the case $N = 2$ that the inner sum to the right in (4.3) contains not more than two terms. Therefore, applying the inequality $|a+b|^2 \leq 2(|a|+|b|)^2$, we obtain

$$\sum_{l \neq 0} \left| \sum_{\substack{n-m=l \\ |n|^2=|m|^2=k}} c_n \bar{c}_m \right|^2 \leq 2 \sum_{|n|^2=k} \sum_{|m|^2=k} |c_n \bar{c}_m|^2 = 2 \left(\sum_{|n|^2=k} |c_n|^2 \right)^2.$$

From this inequality and (4.3) we deduce the desired estimate

$$\|A_k\|^2_{L_4(\mathbb{T}^2)} \leq \frac{\sqrt{3}}{2\pi} \|A_k\|^2_{L_2(\mathbb{T}^2)}. \tag{4.4}$$

This estimate shows that the norms of the "spherical" sums (4.1) in the spaces $L_p(\mathbb{T}^2)$ for $2 \leq p \leq 4$ have the same order of growth.

Cooke's theorem follows in an obvious way from (4.4). Indeed, set

$$E_k(\varepsilon) = \{x \in \mathbb{T}^2 : |A_k(x)| > \varepsilon\}.$$

Then, applying the Cauchy-Bunyakovskiĭ inequality, we obtain

$$\begin{aligned} \|A_k\|^2_{L_2(\mathbb{T}^2)} &\leq |E_k(\varepsilon)|^{1/2} \cdot \|A_k\|^2_{L_4(E_k^2)} + \varepsilon^2 |\mathbb{T}^2 \backslash E_k(\varepsilon)| \\ &\leq |E_k(\varepsilon)|^{1/2} \|A_k\|^2_{L_4(\mathbb{T}^2)} + 4\pi^2 \varepsilon^2. \end{aligned}$$

It follows from this estimate and (4.4) that

$$\|A_k\|^2_{L_4(\mathbb{T}^2)} \leq 4\pi^2\varepsilon^2(2\pi - \sqrt{3}|E_k(\varepsilon)|^{1/2})^{-1}\sqrt{3}$$

for $|E_k(\varepsilon)| < \frac{1}{3}4\pi^2$. If $\lim\limits_{k\to\infty} A_k(x) = 0$ a. e. on $\mathbb{T}^2$ then $|E_k(\varepsilon)| \to 0$ for $k \to \infty$ and any $\varepsilon > 0$. Therefore

$$\overline{\lim_{k\to\infty}} \|A_k\|^2_{L_4(\mathbb{T}^2)} \leq 2\pi\sqrt{3}\cdot\varepsilon^2,$$

so that, as ε was arbitrary,

$$\lim_{k\to\infty} \|A_k\|^2_{L_4(\mathbb{T}^2)} = 0. \tag{4.5}$$

It is clear that the desired result (4.2) follows from (4.5) (and the Cauchy-Bunyakovskiĭ inequality).

Cooke's theorem was the last missing link in the solution of the problem of the unique representation of functions by multiple trigonometric series (cf. Sect. 4 of the introduction). Before this it was proved in Shapiro (1967) that from the convergence to zero of the circular partial sums of a double trigonometric series

$$\sum c_n e^{inx}$$

at each point of $x \in \mathbb{T}^2$ it follows, in the additional assumption of (4.2), that $c_n = 0$ for all $n \in \mathbb{Z}^N$. Cooke's theorem shows that the validity of condition (4.2) is guaranteed by the condition of convergence at each point.

Moreover, as has been shown by Zygmund, for the relation (4.2) to hold true it is sufficient to require that (4.1) holds on a set of positive measure (Zygmund (1972)).

In order to solve the uniqueness problem by analogous methods in the case $N > 2$ one has to extend the theorems of Shapiro and Cooke to N-fold series. So far only the following result by Connes (1976) is available: if the spherical partial sums of the series

$$\sum_{n\in\mathbb{Z}^N} c_n e^{inx} \tag{4.6}$$

converge in some nonempty open set $\omega \subset \mathbb{T}^N$, $N \geq 2$, then the relation (4.2) is fulfilled. However, in the case $N > 2$ this result does not suffice to establish the uniqueness.

§2. The Denjoy-Luzin Theorem

2.1. The Case $N = 1$. If

$$\sum(|a_n| + |b_n|) < \infty, \tag{4.7}$$

then the corresponding trigonometric series converges absolutely and uniformly, so in particular

$$\sum_{n=1}^{\infty} |a_n \cos nx + b_n \sin nx| < \infty. \tag{4.8}$$

Is the converse true, i. e. when does (4.7) follow from (4.8)? An answer to this question is provided by the *theorem of Denjoy-Luzin*: if (4.8) holds on a set of positive measure, then (4.7) is fulfilled. In other words, the series (4.8) either converges everywhere or diverges a. e.

2.2. The Spherical Mean. For the series (4.6) the natural analogue of (4.8) looks as follows:

$$\sum_{k=0}^{\infty} \left| \sum_{|n|^2=k} c_n e^{inx} \right| < \infty. \tag{4.9}$$

Concerning condition (4.7), the affair is somewhat more complicated. One might think that the most immediate counterpart is

$$\sum_{n \in \mathbf{Z}^N} |c_n| < \infty, \tag{4.10}$$

but in the case studied here this is not so. In V. S. Panfërov (1975) it is proved that there exists a double trigonometric series (4.6), $N = 2$, for which condition (4.9) is fulfilled a. e. on $\mathbb{T}^2$, but

$$\sum |c_n| = \infty.$$

In the same paper (Panfërov (1975)) it is shown that for a square double series the natural analogue of (4.7) is the condition

$$\sum_{k=0}^{\infty} \left(\sum_{|n|^2=k} |c_n|^2 \right)^{1/2} < \infty. \tag{4.11}$$

Namely, if (4.9) holds on a set of positive measure on $\mathbb{T}^2$ then (4.11) is fulfilled. Also the converse is true: if (4.11) is fulfilled, then the series (4.9) converges a. e. on $\mathbb{T}^2$. Consequently, for $N = 2$ the series (4.9) either converges a. e. or diverges a. e..

The second half of V. S. Panfërov's result is valid for all $N \geq 2$ and follows immediately from Levi's theorem to the effect that the convergence of a series obtained by integrating term by term implies the convergence a. e. of the function series itself. Indeed, appplying the Cauchy-Bunyakovskiĭ inequality to the integral, we obtain

$$\sum_{k=0}^{\infty}\int_{\mathbb{T}^N}\left|\sum_{|n|^2=k}c_n e^{inx}\right|dx \le \sum_{k=0}^{\infty}|\mathbb{T}^N|^{1/2}\cdot\left(\int_{\mathbb{T}^N}\left|\sum_{|n|^2=k}c_n e^{inx}\right|^2 dx\right)^{1/2}$$
$$= |\mathbb{T}^N|\sum_{k=0}^{\infty}\left(\sum_{|n|^2=k}|c_n|^2\right)^{1/2},$$

and from the convergence of the last series (4.9) follows for a. e. $x \in \mathbb{T}^N$.

The coefficient condition (4.11) can be interpreted in terms of the smoothness of the function which is expanded into the Fourier series (4.6). We state one such result due to S. P. Konovalov (1979). He showed that if $l > 1$ then for each function $f \in C^l(\mathbb{T}^N)$ the series (4.9) converges a. e. on $\mathbb{T}^N$. For $l = 1$ this result is not true, as in Konovalov (1979) there is constructed a function $f \in C^1(\mathbb{T}^N)$, $N \ge 2$, such that the series (4.9) diverges on a set of positive measure.

Combining the result of S. P. Konovalov with the theorem of V. S. Panfërov we find that the Fourier coefficients of any function $f \in C^l(\mathbb{T}^N)$ with $l > 1$ satisfy condition (4.11), while for $l = 1$ and $N = 2$ this is not the case.

In Konovalov (1979) the non trivial part is the construction of a function $f \in C^1(\mathbb{T}^N)$, $N \ge 2$, whose Fourier series (4.6) is not absolutely convergent. Concerning the positive result on convergence in $C^l(\mathbb{T}^N)$ for $l > 1$, the latter can be obtained as an easy consequence of condition (4.11). Indeed, if $f \in C^l(\mathbb{T}^N)$, $l > 1$, is given by the Fourier expansion (4.6), then applying the Cauchy-Bunyakovskiĭ inequality we get

$$\sum_{k=0}^{\infty}\left(\sum_{|n|^2=k}|c_n|^2\right)^{1/2} = \sum_{k=0}^{\infty}(1+k)^{-l/2}\cdot\left(\sum_{|n|^2=k}|c_n|^2(1+|n|^2)^l\right)^{1/2}$$
$$\le \left[\sum_{k=0}^{\infty}(1+k)^{-l}\right]^{1/2}\cdot\left[\sum_{n\in\mathbb{Z}^N}|c_n|^2(1+|n|^2)^l\right]^{1/2}$$
$$\le \text{const }\|f\|_{L_2^l(\mathbb{T}^N)},$$

where $L_2^l(\mathbb{T}^N)$ is the Liouville class defined in Sect. 10 of the introduction (cf. (58)). As $C^l(\mathbb{T}^N) \subset L_2^{l-\varepsilon}(\mathbb{T}^N)$, $\varepsilon > 0$, this implies the required relation (4.11).

2.3. Rectangular Sums. In this section we give another analogue of the Denjoy-Luzin theorem, contained in the paper Revez and Szasz (1957), pertaining to rectangular absolute convergence. Let $\mathbb{Z}_+^N$ stand for the set of elements of the lattice of integers $\mathbb{Z}^N$ with nonnegative coordinates, and set for any $m \in \mathbb{Z}_+^N$

$$\Lambda_m = \{n \in \mathbb{Z}^N : |n_j| = m_j, \quad j = 1, \dots, N\}.$$

In Revez and Szasz (1957) it is shown that if $N = 2$ and

$$\sum_{m\in\mathbf{Z}_+^N} \left| \sum_{n\in\Lambda_m} c_n e^{inx} \right| < \infty$$

on the set of positive measure on $\mathbb{T}^2$ then

$$\sum_{m\in\mathbf{Z}_+^N} \left(\sum_{n\in\Lambda_m} |c_n|^2 \right)^{1/2} < \infty.$$

We note that the converse is trivially true.

§3. Absolute Convergence of Fourier Series

3.1. Some One-Dimensional Results. In this section we consider "real" absolute convergence of Fourier series:

$$f(x) \sim \sum_{n\in\mathbf{Z}^N} f_n e^{inx}, \tag{4.12}$$

i. e. the convergence of the series of its moduli:

$$\sum_{n\in\mathbf{Z}^N} |f_n| < \infty. \tag{4.13}$$

In the one-dimensional case there is a classical result due to S.N. Bernshteĭn (cf. Zygmund (1968), Vol. 1, Chapter VI, (3.1)) on the absolute convergence of Fourier series of functions in the Hölder classes $C^l(\mathbb{T}^N)$: if $l > 1/2$ then the Fourier series is absolutely convergent, while there is a function in $C^{1/2}(\mathbb{T}^N)$ whose Fourier series diverges absolutely.

There is a generalization of this due to Szasz (cf. Zygmund (1968), Vol. 1, Chapter VI, (3.10)): if $C^l(\mathbb{T}^1)$ then

$$\sum |f_n|^\beta < \infty \tag{4.14}$$

for any $\beta > \frac{2}{1+2l}$. If $\beta = \frac{2}{1+2l}$, $l > 0$, it may happen that the series (4.14) is divergent, which is seen at the hand of an example due to Hardy and Littlewood (cf. Zygmund (1968), Vol. 1, Chapter V, (4.2)):

$$f(x) = \sum_{n=1}^{\infty} n^{-\frac{1}{2}-l} e^{in\ln n} e^{inx}. \tag{4.15}$$

The function (4.15) belongs to $C^l(\mathbb{T}^N)$ for any l, $0 < l < 1$, but for it the series (4.14) is divergent if $\beta = \frac{2}{1+2l}$.

3.2. Absolute Convergence of Multiple Series. In the multivariate case one of the conditions for the convergence of (4.13) is easy to obtain. Comparison with a multiple integral shows that the series

$$\sum_{n\in\mathbf{Z}^N} (1+|n|^2)^{-l}$$

is convergent if $l > N/2$. Taking account of this, we can estimate the series (4.13) as follows using the Cauchy-Bunyakovskiĭ inequality:

$$\sum |f_n| = \sum (1+|n|^2)^{-l/2}|f_n| \cdot (1+|n|^2)^{l/2}$$
$$\leq \left(\sum (1+|n|^2)^{-l}\right)^{1/2} \left(\sum |f_n|^2(1+|n|^2)^l\right)^{1/2} = \text{const} \cdot \|f\|_{L_2^l(\mathbb{T}^N)}$$

(for the definition of the norm, see (58) in the introduction). This means that the Fourier series (4.12) of any function $f \in L_2^l(\mathbb{T}^N)$ for $l > N/2$ is absolutely convergent.

From the imbedding $C^l(\mathbb{T}^N) \to L_2^{l-\varepsilon}(\mathbb{T}^N)$, $\varepsilon > 0$, follows now the analogue of Bernshteĭn's theorem: the Fourier series of a function $f \in C^l(\mathbb{T}^N)$ with $l > N/2$ is absolutely convergent.

How sharp is this result for the classes $L_2^l(\mathbb{T}^N)$? In order to answer this question let us consider the numerical series

$$\sum_{n\in\mathbf{Z}^N} (1+|n|^2)^{-\frac{N}{2}}[\log(1+|n|^2)]^{-\alpha},$$

which by Cauchy's convergence test converges if $\alpha > 1$ and diverges if $\alpha \leq 1$. By the Riesz-Fischer theorem the function

$$g(x) = \sum_{n\in\mathbf{Z}^N} (1+|n|^2)^{-\frac{N}{4}}[\log(1+|n|^2)]^{-1}e^{inx} \tag{4.16}$$

is in $L_2(\mathbb{T}^N)$ so that the function

$$f(x) = \sum_{n\in\mathbf{Z}^N} (1+|n|^2)^{-\frac{N}{4}} g_n e^{inx}, \tag{4.17}$$

where g_n are the Fourier coefficients of g, belongs to the Liouville class $L_2^{N/2}(\mathbb{T}^N)$ (cf. Sect. 10 of the introduction). However, it is manifest that the Fourier coefficients of f do not satisfy (4.13). Thus, the condition $l > N/2$ for the absolute convergence of Fourier series in $L_2^l(\mathbb{T}^N)$ is sharp.

The question of absolute convergence in the classes L_p^l is more difficult. If $1 < p < 2$ then the imbedding theorem (cf. (57) in the introduction)

$$L_p^{N/p+\varepsilon}(\mathbb{T}^N) \to L_2^{N/2+\varepsilon}(\mathbb{T}^N)$$

guarantees that (4.13) holds for any $f \in L_p^l(\mathbb{T}^N)$ for $l > N/p$ and, as examples show, this condition is sharp, because the opposite condition $l \leq N/p$ entails that there exists such a function $f(x)$ which is unbounded. If the Fourier coefficients of this function would satisfy (4.13) then this would imply that the series (4.12) converges uniformly to f, which again contradicts the unboundedness of f.

The case $p > 2$ is different. Then we have the imbedding

$$L_p^l \to L_2^l, \quad p > 2,$$

where the smoothness exponent l does not change. Therefore, for $p > 2$ the sufficient condition for absolute convergence of the Fourier series takes the same form as for $p = 2$, i. e. $l > N/2$. The fact that it is not possible to improve this condition was first established by V. A. Il'in (1958), using special properties of Rademacher functions.

Recall that the *Rademacher functions*

$$\rho_k(t) = \operatorname{sign}\sin(2^{k+1}\pi t), \quad k = 0, 1, \ldots, \; t \in [0, 1]$$

form an orthonormal system (not complete!) on $[0, 1]$. We require here its N-dimensional analogue: we set

$$r_k(t) = \begin{cases} \rho_{2k}(t), & k \geq 0 \\ \rho_{-2k-1}(t), & k < 0, \end{cases}$$

and consider the functions

$$r_n(t) = r_{n_1}(t_1) \ldots r_{n_N}(t_N), \quad n \in \mathbb{Z}^N, \quad t \in Q,$$

defined in the cube $Q = \{t \in \mathbb{R}^N, 0 \leq t_j \leq 1, \; j = 1, 2, \ldots, N\}$ and likewise forming an orthonormal system. One of the most remarkable properties of this system is the fact that if

$$F(t) = \sum_{n \in \mathbb{Z}^N} c_n r_n(t),$$

then for any p, $1 < p < \infty$, the following inequality holds:

$$A_p \|F\|_{L_2(Q)} \leq \|F\|_{L_p(Q)} \leq B_p \|F\|_{L_2(Q)} \tag{4.18}$$

(cf. Stein (1971), Appendix C). Let g be the function defined by (4.16). Set

$$F(x,t) = \sum_{n \in \mathbf{Z}^N} g_n r_n(t) e^{inx}, \quad x \in \mathbb{T}^N, \quad t \in Q, \tag{4.19}$$

where g_n are the Fourier coefficients of g. It follows from the orthogonality of the Rademacher system that

$$\|F(x,t)\|^2_{L_2(Q)} = \sum_{|n| \in \mathbf{Z}^N} |g_n|^2 = (2\pi)^{-N} \|g\|^2_{L_2(\mathbf{T}^N)} .$$

Taking account of this relation we get from the rightmost inequality (4.18)

$$\int_Q |F(x,t)|^p dt \leq B_p^p \cdot (2\pi)^{-Np} \cdot \|g\|^p_{L_2(\mathbf{T}^N)} .$$

Consequently,

$$\int_{\mathbf{T}^N} dx \int_Q |F(x,t)|^p dt < \infty.$$

Now we can refer to Fubini's theorem, according to which for a. e. $t \in Q$ holds

$$\int_{\mathbf{T}^N} |F(x,t)|^p dx < \infty,$$

i. e. (4.19) (considered as a function of x) belongs to $L_p(\mathbb{T}^N)$ for a. e. t. But then, in view of the definition of Liouville classes (cf. introduction, Sect. 10), viewed as a function of x

$$f(x,t) = \sum_{n \in \mathbf{Z}^N} r_n(t)(1+|n|^2)^{-\frac{N}{4}} g_n e^{inx} \tag{4.20}$$

belongs for a. e. x to $L_p^{N/2}(\mathbb{T}^N)$. If $t \in Q$ has dyadic-irrational coordinates, then $|r_k(t)| = 1$ and for such t the series (4.20) and (4.17) are simultaneously absolutely convergent or absolutely divergent. As (4.17) is absolutely divergent, it follows that (4.20) must be absolutely divergent for a. e. t. Consequently, for each p, $1 < p < \infty$, there exists a function in $L_p^{N/2}(\mathbb{T}^N)$ with absolutely divergent Fourier series.

Let us remark that the above proof shows more: that for each function $f \in L_2^l(\mathbb{T}^N)$ with Fourier series (4.12) "almost all" functions of the form

$$\tilde{f}(x) \sim \sum_{N \in \mathbf{Z}^N} \pm f_n e^{inx},$$

obtained by applying an arbitrary permutation of signs, belong to $L_p^l(\mathbb{T}^N)$, $1 < p < \infty$. Hence, taking $f \in L_2^{N/2}(\mathbb{T}^N)$ with absolutely divergent Fourier series we obtain the requested counterexample.

Thus, in the classes $L_p^l(\mathbb{T}^N)$ the sharp conditions for absolute convergence of Fourier series take the form:

$$l > N \cdot \max\left\{\frac{1}{p}, \frac{1}{2}\right\}, \quad 1 < p < \infty.$$

If $l = N \cdot \max\{\frac{1}{p}, \frac{1}{2}\}$ then there exist $f \in L_p^l(\mathbb{T}^N)$ with an absolutely divergent Fourier series.

3.3. On the Convergence of the Series of Powers of $|f_n|$. For multiple Fourier series one further studies the question which smoothness of the function f will guarantee the convergence of the series

$$\sum_{n \in \mathbf{Z}^N} |f_n|^\beta. \tag{4.21}$$

In Szasz and Minakshisundaram (1947) it is proved that for every function $f \in C^l(\mathbb{T}^N)$ the series (4.21) converges for $\beta > 2N/(N+2l)$. In that paper it is assumed that $0 < l < 1$, which excludes (for $N \geq 2$) the case $\beta = 1$. However, Bochner (1947) has remarked that the result remains in force for any $l > 0$. In particular, if $l > N/2$ this implies the convergence of the series (4.13), and this statement, as we observed in the previous section, is an analogue of Bernshteĭn's theorem.

That the condition $\beta > 2N/(N+2l)$ is sharp follows easily by the example of a series considered by Wainger (1965)

$$f(x) = \sum_{|n|>1} |n|^{-N/2-l} (\log|n|)^{-\frac{N+l}{2N}} e^{i|n|(\log|n|)^d} e^{inx} \tag{4.22}$$

(compare with the Hardy-Littlewood series (4.15)). In Wainger (1965) it is proved that for any $d > 0$ and $l > d^2 N$ the function $f(x)$ defined by (4.22) belongs to $C^l(\mathbb{T}^N)$. The corresponding series (4.21) for $\beta = 2N/(N+2l)$ has the form

$$\sum_{n \in \mathbf{Z}^N} |f_n|^{\frac{2N}{N+2l}} = \sum_{|n|>1} |n|^{-N} (\log|n|)^{-\frac{N+l}{N+2l}}$$

and, clearly, is divergent for every $l > 0$.

Remarks and Bibliographical Notes

Multiple Fourier series are treated in the books Zhizhiashvili (1969), Yanushauskas (1986), Stein and Weiss (1971), Chapter VIII and the surveys articles Alimov, Il'in, and Nikishin (1976/77), Zhizhiashvili (1973), Ash (1976). In the bibliographical survey Golubov (1982) most results in the theory of multiple Fourier series and Fourier integrals, reviewed in Referativnyĭ Zhurnal "Matematika" between the years 1953 and 1980, are set forth (and in some cases likewise their generalization to eigenfunctions and to spectral expansions is mentioned).

The theory of general orthogonal series is developed in the book Kaczmarz and Steinhaus (1951).

Fourier series expansions in eigenfunctions of elliptic operators and also the corresponding spectral expansions are treated in Titchmarsh (1958) and in the expository papers Alimov, Il'in, and Nikishin (1976/77), Il'in (1958), (1968). Moreover, the papers Il'in and Alimov (1971), Hörmander (1968), (1969) contain a short survey of results and methods, along with a bibliography.

We remark that many results on multiple Fourier series and Fourier integrals, mentioned in the present survey, have been proved in a more general situation (for example, for Fourier series expansions for Laplace operator or general elliptic operators). The results of the following papers belong to this category: Alimov (1973), (1974), (1976), (1978), Alimov, Il'in, and Nikishin (1976/77), Ashurov (1983b), Babenko (1973b), (1978), Il'in (1957), (1958), (1968), Il'in and Alimov (1971), (1972), Mityagin and Nikishin (1973), Nikishin (1972b), Pulatov (1978), Hörmander (1968), (1969), Peetre (1964), Titchmarsh (1958).

In the present survey we have not discussed more general questions of the theory of multiple Fourier series (for example, Fourier series on groups). These questions are discussed in detail in the introductory article by V. P. Khavin (1987) in Vol. 15 "Commutative Harmonic Analysis I" of this Encyclopaedia (cf. further other parts of the series "Commutative Harmonic Analysis"). There one finds also a bibliography and a detailed account of the historical development of harmonic analysis, in particular, of the theory of Fourier series.

The majority of the results in the one-dimensional theory of trigonometric series can be found in Bari (1961), Zygmund (1968). Moreover, we mention the survey article Kislyakov (1988) in this series.

Introduction. Sects. 3 and 4. Concerning results pertaining to partial sums and forms of convergence, not considered in the present work, we refer to the papers Golubov (1982), Ash (1976). Note that the hyperbolic partial sums (cf., for example, Golubov (1982), Belinskiĭ (1979)) $\sum_{|n_1|\cdot\ldots\cdot|n_N|\le R} c_n e^{inx}$ do not fall within the framework of Definition D) in Sect. 3, because the set $\{n \in \mathbb{Z}^N : |n_1|\cdot\ldots\cdot|n_N| \le R\}$ is not bounded. Such partial sums play a great rôle in approximation theory.

Sect. 6. In this survey we consider only Riesz means. Other summation methods (cf. Abel, Riemann and Poisson means) are considered e. g. in Alimov, Il'in, and Nikishin (1976/77), Trigub (1980), Yanushauskas (1986), Stein and Weiss (1971), Chapter VII, Zygmund (1968), Vol. II. In particular, Abel means of multiple Fourier series are used in the study of the boundary behavior of analytic functions of several complex variables (cf., for example, Zygmund (1968), Chapter XVII, Stein and Weiss (1971), Yanushauskas (1986)).

Sect. 8. The asymptotic behavior of the Fourier transform of characteristic functions of convex sets is studied in Herz (1962), Randol (1969).

Section 9. In the study of the convergence of multiple Fourier series further the asymptotic behavior for large $R > 0$ of the Lebesgue constant

$$L_R(\omega) = \int_{\mathbb{T}^N} \left| \sum_{n\in R\omega\cap\mathbb{Z}^N} e^{inx} \right| dx$$

(i. e. the norm of the operation of taking the partial sum $S_{R\omega}(x, f)$ in $C(\mathbb{T}^N)$) plays an important rôle. The following two-sided estimate (Yudin (1979)) holds true: $c_1 \ln^N R \le L_R(\omega) \le c_2 R^{(N-1)/2}$. Hence, within the whole class of convex sets the Lebesgue constant of the ball has the largest growth (cf. (1.19)), while the Lebesgue constant of the circle has the smallest growth. For $N = 2$ A. N. Podkorytov (1984) has constructed for any given numbers $p > 2$ and $0 < q < \frac{1}{2}$ convex sets $\omega = \omega(p)$ and $\sigma = \sigma(q)$ such that $c_1 \ln^p R \le L_R(\omega) \le c_2 \ln^p R$ and $c_1 R^q \le L_R(\sigma) \le c_2 R^q$.

Let us remark that the left hand side of (40) equals $N(\mu)$, the number of eigenvalues not exceeding μ. In the study of the problem of the circle in number theory one uses in the multidimensional case effectively the method of trigonometric sums (cf., for example, Hua (1959); note that in this paper, as well as in Babenko (1978), likewise estimates of the type (39) are obtained).

Chapter 1. Sect. 1.2. In Goffman and Liu (1972) it is further proved that the localization principle holds in the class $W_1^{N-1}(\mathbb{T}^N)$ under square summation.

Sect. 1.4. The results on equiconvergence of multiple Fourier series and Fourier integrals are due to Bochner (1936). Stein (1958) sharpened Bochner's result on localization for $s = \frac{N-1}{2}$ in terms of the Orlicz class $L\log^+ L$ ($f \in L\log^+ L$ if and only if the function $|f(x)|\log(\max\{1, |f(x)|\})$ is summable).

Sect. 1.5. B. M. Levitan (1963) gave necessary conditions for the localization in the classes W_2^l. For odd N these conditions coincide with (1.18) and are of final character. An analogue of V. A. Il'in's theorem (on the failure of localization for f in C^l) in the Nikol'skiĭ classes H_∞^l was obtained in Pulatov (1978).

Sect. 1.6. That the condition (1.25) is sharp for multiple Fourier integrals within the class of all elliptic polynomials was established in Pulatov (1978). In A. Ĭ. Bastis (1982) there is constructed a selfadjoint extension of the Laplace operator such that for the eigenfunction expansion condition (1.25) cannot be improved for $p = 1$.

Sect. 2.2. If $f \in H_p^l(\mathbb{T}^N)$, $pl > N$, then by the imbedding theorem $f \in C^\varepsilon(\mathbb{T}^N)$, $\varepsilon = l - \frac{N}{p}$ (cf. (59)), so that for such values of p and l the condition of uniform convergence (1.3) can not be fulfilled.

Sect. 2.3. Stein (1958) proved that uniform summability of Riesz means of the critical order holds likewise for functions in the Orlicz class $L\log^+ L$. In the classes of l times continuously differentiable functions a condition for uniform convergence was obtained by B. M. Levitan (1963). For odd l this result is sharp.

Sect. 2.4. If the function $f \in L_p(G)$, $1 \leq p \leq 2$, is continuous in a subdomain $D \subset G$, then the Riesz means of order $s > \frac{N-1}{p}$ of the spectral expansion of the function, corresponding to an arbitrary elliptic operator, converge to $f(x)$ uniformly on every compact set $K \subset D$. This result is due to Hörmander (1969).

Chapter 2. Sect. 1.1. L. V. Zhizhiashvili (1970), (1971), (1973) obtained a definitive condition for convergence in $L_1(\mathbb{T}^N)$ for functions in $L_1(\mathbb{T}^N)$ in terms of an integral modulus of continuity.

Sect. 1.2. The convergence in L_q of spectral expansions of functions in L_p is studied in Mityagin and Nikishin (1973a), (1973b).

Sect. 1.3. In K. I. Babenko (1973a), (1973b) the order of growth in λ of the norm of the operator $E_\lambda^s : L_p \to L_q$ is studied in the assumption of (2.10). Condition (2.9) for $p = 1$ was sharpened in Orlicz classes by Stein (1958).

Sect. 3. The theory of multipliers and its applications is set forth in the books Nikol'skiĭ (1977), Hörmander (1983–85), Stein (1970), Stein and Weiss (1971), Zygmund (1968). Pseudo differential operators and Fourier integral operators play a major rôle in the theory of boundary problems of mathematical physics (cf., e. g., Maslov (1965), Hörmander (1983–85), Taylor (1981)).

Sect. 3.3. If the characteristic function of a set $E \subset \mathbb{R}^N$ is a multiplier on $L_p(\mathbb{R}^N)$, then the same property is enjoyed by any set E' obtained by rotating E. In the periodic case this property may fail (cf. Belinskiĭ (1979)).

Sect. 3.5. If the elliptic partial differential operator $A(x, D)$ has a complete set of eigenfunctions $\{u_n(x)\}$ in $L_2(\Omega)$, then the partial sum of the Fourier series with respect to this system can be written as an integral operator:

$$E_\lambda f(x) = \int_\Omega \Theta(x, y, \lambda) f(y)\, dy, \quad f \in L_2(\Omega).$$

The kernel $\Theta(x, y, \lambda)$ is called the spectral function of the operator $A(x, D)$ and takes the form

$$\Theta(x, y, \lambda) = \sum_{\lambda_n < \lambda} u_n(x) u_n(y),$$

where λ_n is the eigenvalue corresponding to u_n. Let $N(\lambda)$ be the number of eigenvalues not exceeding λ. Then

$$N(\lambda) = \int_\Omega \Theta(x, x, \lambda)\, dx.$$

The asymptotic estimate for $N(\lambda)$ goes back to the work of Weyl, Courant and Carleman. For a large class of boundary problems one knows that (cf. Hörmander (1983–85))

$$N(\lambda) \sim (2\pi)^{-N} \int_\Omega \left(\int_{A(x,\xi)<\lambda} d\xi \right) dx.$$

This estimate is a generalization of formula (40).

So far two approaches are known to the study of the asymptotics of the spectral function $\Theta(x, y, \lambda)$, where x and y lie on a compact subset of Ω. To the first group belongs the work of Carleman, B. M. Levitan, Gårding, Hörmander and others, which authors in order to obtain estimates for the remainder term of the spectral function employ the method of Carleman, which amounts to considering a suitable function of the operator A followed by a subsequent application of an appropriate Tauberian theorem. For elliptic second order operators V. A. Il'in (cf. Il'in (1958), (1968), Alimov, Il'in and Nikishin (1976/77), Part II) has developed an alternative approach, which is based only on the mean value theorem. Let us here mention the most general result, due to Hörmander:

$$\Theta(x, y, \lambda) = (2\pi)^{-N} \int_{A(x,\xi)<\lambda} e^{i\psi(x,y,\xi)} d\xi + O(\lambda^{\frac{N-1}{m}}),$$

where m is the order of A and $\psi(x, y, \lambda)$ a phase function (cf. (2.53)).

Chapter 3. Sect. 1.2. In Bakhbukh (1974), Oskolkov (1974) conditions are given for the a. e. convergence of a multiple Fourier series of a continuous function in terms of the modulus of continuity. These conditions are almost definitive (Bakhbukh and Nikishin (1974)). Analogous results in terms of integral moduli of continuity in the multidimensional situation are given in Zhizhiashvili (1971), (1973).

Sect. 2.1. When $f \in L_2(\mathbb{T}^N)$ is "logarithmically smooth" in terms of an integral modulus of continuity, it has been shown by V. I. Golubov (1975) that $E_\lambda f(x) \to f(x)$ a. e. The a. e. convergence of the Riesz means $E_\lambda^s f(x)$ of critical order $s = \frac{N-1}{2}$ for $f \in L(\log^+ L)^2$ was proven by Stein (1958), (1961). In a recent publication Sunouchi (1985) has shown that this result extends also to functions $f \in L \log^+ L \log^+ \log^+ L$.

Sect. 2.2. Divergence on a set of positive measure is also studied in Mityagin and Nikishin (1973a). Let us remark that the condition (3.24) for a. e. convergence of multiple Fourier series and integrals, in contrast to localization, does not depend on the geometry of the surface ∂Q_A (Ashurov (1983b)).

Chapter 4. Sect. 2.2. The Denjoy-Luzin theorem for $N = 2$ is also proved in Golubov (1975).

Sect. 3.2. In several papers necessary conditions for the absolute convergence of multiple Fourier integrals are studied. Let us mention R. M. Trigub (1980).

Bibliography*

Alimov, Sh. A. (1973): Uniform convergence and summability of spectral expansions of functions in L_a^p. Differ. Uravn. *9*, No. 4, 669–681. English translation: Differ. Equations *9*, 509–517 (1975), Zbl. 259.35064.

Alimov, Sh. A. (1974): On the localization of spectral decompositions. Differ. Uravn. *10*, No. 4, 744–746. English translation: Differ. Equations *10*, 576–578 (1975), Zbl. 284.35059.

Alimov, Sh. A. (1976): On spectral decompositions of functions in H_p^a. Mat. Sb., Nov. Ser. *101*, No. 1, 3–21. English translation: Math. USSR, Sb. *30*, 1–16 (1978), Zbl. 344.46069.

Alimov, Sh. A. (1978): On expanding continuous functions from Sobolev classes in eigenfunctions of Laplace operator. Sib. Math. Zh. *19*, No. 1, 721–734. English translation: Sib. Math. J. *19*, 507–517 (1979), Zbl. 397.47023.

Alimov, Sh. A., Il'in, V. A., and Nikishin, E. M. (1976/77): Questions of convergence of multiple trigonometric series and spectral decompositions. Usp. Mat. Nauk *31*, No. 6, 28–83; *32*, No. 1, 107–130. English translation: Russ. Math. Surv. *31*, No. 6, 29–86 (1976); *32*, No. 1, 115–139 (1977), Zbl. 345.42002; Zbl. 353.42005.

Ash, J. M. (1976): Multiple trigonometric series. In: Stud. harmon. Anal. Proc Conf. Chicago 1974, MAA Stud. Math. 13, 76–96. Zbl. 358.42011.

Ashurov, R. R. (1983a): On conditions for localization of spectral expansions corresponding to elliptic operators with constant coefficients. Mat. Zametki *33*, No. 6, 847–856. English translation: Math. Notes *33*, 434–439 (1983), Zbl. 588.35065.

Ashurov, R. R. (1983b): On the a. e. summability of Fourier series in L_p with respect to eigenfunctions. Mat. Zametki *34*, No. 6, 837–843. English translation: Math. Notes *34*, 913–916 (1983), Zbl. 536.42029.

Ashurov, R. R. (1985): On localization conditions for multiple trigonometric Fourier series. Dokl. Akad. Nauk SSSR *282*, No. 4, 777–780. English translation: Sov. Math., Dokl. *31*, 496–499 (1985), Zbl. 613.42009.

Babenko, K. I. (1973a): On the mean convergence of multiple Fourier series and the asymptotics of the Dirichlet kernel for spherical means. Preprint No. 52. Moscow: IPM Akad. Nauk SSSR. [Russian]

Babenko, K. I. (1973b): On the summability and convergence of eigenfunction expansions of a differential operator. Mat. Sb., Nov. Ser. *91*, No. 2, 147–201. English translation: Math. USSR, Sb. *20*, 157–211 (1974), Zbl. 283.35060.

Babenko, K. I. (1978): On the asymptotics of the Dirichlet-kernel of spherical means of multiple Fourier series. Dokl. Akad. Nauk SSSR *243*, No. 5, 1097–1100. English translation: Sov. Math., Dokl. *19*,1457–1461 (1978), Zbl. 419.10046.

Bakhbukh, M. (1974): On sufficient conditions for rectangular convergence of double series. Mat. Zametki *15*, No. 6, 835–838. English translation: Math. Notes *15*, 501–503 (1974), Zbl. 314.42011.

Bakhbukh, M. and Nikishin, E. M. (1974): On the convergence of double Fourier series of continuous functions. Sib. Math. Zh. *14*, No. 6, 1189–1199. English translation: Sib. Math. J. *14*, 832–839 (1974), Zbl. 279.42021.

Bari, N. K. (1961): Trigonometric series. Moscow: Fizmatgiz. English translation: Oxford: Pergamon Press 1964 (2 vols.)

Bastis, A. Ĭ. (1982): On the divergence of certain spectral expansions in L_1^a. Mat. Zametki *32*, No. 3, 309–314. English translation: Math. Notes *32*, 634–637 (1983), Zbl. 504.35069.

* For the convenience of the reader, references to reviews in Zentralblatt für Mathematik (Zbl.), compiled using the MATH database, and Jahrbuch über die Fortschritte der Mathematik (Jbuch) have, as far as possible, been included in this bibliography.

Bastis, A. Ĭ. (1983): Some questions of the summability of expansions corresponding to elliptic operators. Abstract of candidate dissertation. Moscow: MGU. [Russian]

Bastis, A. Y. (1983): On the asymptotics of the Riesz-Bochner kernel. Annal. Math. *9*, 247–258, Zbl. 543.42012.

Belinskiĭ, È. S. (1979): Some properties of hyperbolic partial sums. In: Theory of functions and mappings. Collect. Sci. Works, Kiev 1979, pp. 28–36, [Russian] Zbl. 493.42021.

Bloshanskiĭ, I. L. (1975): On the uniform convergence of multiple trigonometric series and Fourier integrals, Mat. Zametki *18*, No. 2, 153–168. English translation: Math. Notes *18*, 675–684 (1976), Zbl. 317.42018.

Bloshanskiĭ, I. L. (1976): On the uniform convergence of multiple trigonometric series and Fourier integrals under square summation. Izv. Akad. Nauk SSSR, Ser. Mat. *40*, 685–705. English translation: Math. USSR, Izv. *10*, 652–671 (1977), Zbl. 348.42005.

Bochner, S. (1936): Summation of multiple Fourier series by spherical means. Trans. Am. Math. Soc. *40*, 175–207, Zbl. 15,157.

Bochner, S. (1947): Review of "On absolute convergence of multiple Fourier series" by Szasz and Minakshisundaram. Math. Rev. *8*, 376

Carleson, L. (1966): On convergence and growth of partial sums of Fourier series. Acta Math. *116*, No. 1–2, 135–137, Zbl. 144,64.

Carleson, L. and Sjölin, P. (1972): Oscillatory integrals and multiplier problem for the disc. Stud. Math. *44*, No. 3, 287–299, Zbl. 215,183.

Connes, B. (1976): Sur les coefficients des séries trigonométriques convergentes sphériquement. C. R. Acad. Sci., Paris, Ser. A *283*, No. 4, 159–161, Zbl. 335.42005

Cooke, R. (1971): A Cantor-Lebesgue theorem in two dimensions. Proc. Am. Math. Soc. *30*, No. 4, 547–550, Zbl. 222.42014.

Fefferman, C. (1970): Inequalities for strongly singular integrals. Acta Math. *124*, 9–36, Zbl. 188,426

Fefferman, C. (1970): The multiplier problem for the ball. Ann. Math., II. Ser. *94*, 330–336, Zbl. 234.42009.

Fefferman, C. (1971): On the divergence of multiple Fourier series. Bull. Am. Math. Soc. *77*, No. 2, 191–195, Zbl. 212,94.

Garcia-Cuerva, J. and Rubio de Francia, J. L. (1985): Weigthed norm inequalities and related topics. Amsterdam: North-Holland (604 p.), Zbl. 578.46046.

Goffman, C. and Liu, Fon-Che (1972): On the localization property of square partial sums for multiple Fourier series. Stud. Math. *44*, No. 1, 61–69, Zbl. 241.42016.

Golubov, B. I. (1975): On the convergence of spherical Riesz means of multiple Fourier series. Mat. Sb., Nov. Ser. *96*, 189–211. English translation: Math. USSR, Sb. *25*, (1976), Zbl. 305.42022.

Golubov, B. I. (1977): On the summability of Fourier integrals of spherical Riesz means. Mat. Sb., Nov. Ser. *104*, No. 4, 577–596. English translation: Math. USSR, Sb. *33*, 501–518 (1977), Zbl. 364.42019.

Golubov, B. I. (1982): Multiple Fourier series and Fourier integrals. In: Itogi Nauki Tekh. Ser. Mat. Anal. *19*, 3–54. English translation: J. Sov. Math. *24*, 639–673 (1984), Zbl. 511.42021.

de Guzman, M. (1981): Real variable methods in Fourier analysis. Amsterdam New York Oxford: North-Holland, Zbl. 449.42001.

Herz, C. S. (1962): Fourier transforms related to convex sets. Ann. Math., II. Ser. *75*, No. 2, 81–92, Zbl. 111,348.

Hörmander, L. (1960): Estimates for translation invariant operators in L^p spaces. Acta Math. *104*, 93–140, Zbl. 93,114.

Hörmander, L. (1968): The spectral function of an elliptic operator. Acta Math. *121*, No. 3–4, 193–218, Zbl. 164,132,

Hörmander, L. (1969): On the Riesz means of spectral functions and eigenfunction expansions for elliptic differential operators. In: Some recent advances in the basic sciences,

Vol. 2 (Proc. Annual Sci. Conf., Belfer Grad. School Sci., Yeshiva Univ., New York, 1965–66), pp. 155–202. New York: Belfer Graduate School, Yeshiva Univ.
Hörmander, L. (1983–85): The analysis of linear partial differential operators I–IV. (Grundlehren 256, 257, 274, 275.) Berlin Heidelberg: Springer-Verlag, Zbl. 521.35001, Zbl. 521.35002, Zbl. 601.35001, Zbl. 612.35001.
Hua Loo-Keng (1959): Abschätzungen von Exponentialsummen und ihre Anwendung in der Zahlentheorie. Leipzig: Teubner (169 p.), Zbl. 83,36.
Hunt, R. A. (1968): On convergence of Fourier series, in: Proc. Conf. on Orthogonal Expansions and their Continuous Analogues, Edwardsville, pp. 235–255. Carbondale: Univ. Press, Zbl. 159,357.
Il'in, V. A. (1957): On uniform eigenfunction expansions under summation taken under increasing order of the eigenvalues. Dokl. Akad. Nauk SSSR *114*, No. 4, 698–701, [Russian] Zbl. 79,291.
Il'in, V. A. (1958): On convergence of eigenfunction expansions of the Laplace operator. Usp. Mat. Nauk *13*, No. 1, 87–180, [Russian] Zbl. 79,292.
Il'in, V. A. (1968): The problems of localization and convergence of Fourier series with respect to the fundamental systems of functions of the Laplace operator. Usp. Mat. Nauk *23*, No. 2, 61–120. English translation: Russ. Math. Surv. *23*, No. 2, 59–116 (1968), Zbl. 189,357.
Il'in, V. A. (1970): Localization conditions for rectangular partial sums of the multiple trigonometric Fourier series in the classes of S. M. Nikol'skiĭ. Mat. Zametki *8*, No. 5, 595–606. English translation: Math. Notes *8*, 803–809 (1971), Zbl. 212,95.
Il'in, V. A. and Alimov, Sh. A. (1971): Conditions for the convergence of spectral expansions corresponding to selfadjoint extensions of elliptic operators, I. Differ. Uravn. *7*, No. 4, 670–710. English translation: Differ. Equations *7*, 516–543 (1973), Zbl. 224.35014.
Il'in, V. A. and Alimov, Sh. A. (1972): On the divergence on a set of positive measure of Riesz means of kernels of fractional order. Differ. Uravn. *8*, No. 8, 372–373. English translation: Differ. Equations *8*, 283–284 (1974), Zbl. 244.35033.
Kaczmarz, S. and Steinhaus, H. (1951): Theorie der Orthogonalreihen. New York: Chelsea, Zbl. 45,336.
Khavin, V. P. (1987): Methods and structure of commutative harmonic analysis. In: Itogi Nauki Tekhn., Ser. Sovrem. Probl. Mat. *15*, 6–133. Moscow: VINITI. English translation: Encyclopaedia Math. Sci. Vol. 15, pp. 1–111. Berlin Heidelberg: Springer-Verlag (1991)
Kislyakov, S. V. (1987): Classical themes of Fourier analysis. In: Itogi Nauki Tekhn., Ser. Sovrem. Probl. Mat. *15*, 135–195. Moscow: VINITI. English translation: Encyclopaedia Math. Sci. Vol. 15, pp. 113–165. Berlin Heidelberg: Springer-Verlag (1991), Zbl. 655.42006.
Konovalov, S. P. (1979): On absolute convergence of multiple Fourier series. Mat. Zametki *25*, No. 2, 211–216. English translation: Math. Notes *25*, 109–112 (1979), Zbl. 398.42014.
de Leeuw, K. (1965): On L^p multipliers. Ann. Math., II. Ser. *81*, 364–379, Zbl. 171,118.
Levitan, B. M. (1955): On the summation of multiple Fourier series and Fourier integrals. Dokl. Akad. Nauk SSSR *102*, No. 6, 1073–1076, [Russian] Zbl. 68,279.
Lizorkin, P. I. (1963): Generalization of Liouville differentiation and the function space $L_p^r(E_n)$. Mat. Sb., Nov. Ser. *60*, No. 3, 325–353, [Russian] Zbl. 134,317.
Marcinkiewicz, J. (1939): Sur les multiplicateurs des séries de Fourier. Stud. Math. *8*, 78–91, Zbl. 20,354.
Maslov, V. P. (1965): Perturbation theory and asymptotic methods. Moscow: MGU. French translation: Paris: Dunod, Gauthiers-Villars 1972, Zbl. 247.47010.
Meshkov, V. Z. (1978): On spherical multipliers. Mat. Zametki *23*, No. 1, 105–112 English translation: Math. Notes *23*, 58–62 (1978), Zbl. 387.42005.
Mikhlin, S. G. (1956): On multipliers of Fourier integrals. Dokl. Akad. Nauk SSSR *109*, 701–703, [Russian] Zbl. 73,84.
Mitchell, J. (1951): On the spherical summability of multiple orthogonal series. Trans. Am. Math. Soc. *71*, 136–151, Zbl. 44,286.

Mityagin, B. S. and Nikishin, E. M. (1973a): On a. e. divergence of Fourier series. Dokl. Akad. Nauk SSSR *210*, No. 1, 23–25. English translation: Sov. Math., Dokl. *14*, 677–680 (1973), Zbl. 287.42009.

Mityagin, B. S. and Nikishin, E. M. (1973b): On divergence of spectral expansions in the mean and a. e. Dokl. Akad. Nauk SSSR *212*, No. 3, 551–552. English translation: Sov. Math., Dokl. *14*, 1417–1419 (1973), Zbl. 298.35045.

Nikishin, E. M. (1972a): Weyl multipliers for multiple Fourier series. Mat. Sb., Nov. Ser. *89*, No. 2, 340–348. English translation: Math. USSR, Sb. *18*, 351–360 (1973), Zbl. 258.42029.

Nikishin, E. M. (1972b): A resonance theorem and series in eigenfunctions of the Laplace operator. Izv. Akad. Nauk SSSR, Ser. Mat. *36*, No. 4, 795–813. English translation: Math. USSR, Izv. *6*, 788–806 (1973), Zbl. 258.42018.

Nikol'skiĭ, S. M. (1977): Approximation of functions of several variables and imbedding theorems. 2nd ed. Moscow: Nauka (456 pp.). English translation: New York: Springer-Verlag 1975 (Grundlehren 205), Zbl. 185,379.

Oskolkov K. I. (1974): An estimate for the rate of approximation of a continuous function and its conjugate function by Fourier sums on a set of positive measure. Izv. Akad. Nauk SSSR, Ser. Mat. *38*, No. 6, 1393–1407. English translation: Math. USSR, Izv. *8*, 1372–1386 (1976), Zbl. 307.42002.

Panfërov, V. S. (1975): Analogues of the theorems of Luzin-Denjoy and Cantor-Lebesgue for double trigonometric series. Mat. Zametki *18*, No. 5, 659–674. English translation: Math. Notes *18*, 983–992 (1976), Zbl. 369.42016.

Peetre, J. (1964): Remarks on eigenfunction expansions for elliptic operators with constant coefficients. Math. Scand. *15*, 83–92, Zbl. 131,98.

Podkorytov, A. N. (1984): Intermediate rates of growth of the Lebesgue constants in the two dimensional case. Zap. Nauchn. Semin. Leningr. Otd. Mat. Inst. Steklova *139*, 148–155, [Russian] Zbl. 601.42021.

Pulatov, A. K. (1978): On the localization of spectral expansions connected with homogeneous elliptic operators. Dokl. Akad. Nauk SSSR *248*, No. 4, 800–803. English translation: Sov. Math., Dokl. *19*, 107–110 (1978), Zbl. 393.35048.

Pulatov, A. K. (1981): On conditions for the localization of expansions in N-fold Fourier integrals corresponding to a homogeneous elliptic operator. In: Differential equations and questions of bifurcation theory. Tashkent, 110–123, [Russian] Zbl. 533.47044.

Randol, B. (1969): On the asymptotical behaviour of the Fourier transforms of the indicator functions of a convex set. Trans. Am. Math. Soc. *139*, 279–285, Zbl. 183,269.

Reves, G. and Szasz, O. (1942): Some theorems on double trigonometric series. Duke Math. J. *9*, 693–705, Zbl. 60,191.

Shapiro, V. L. (1957): Uniqueness of multiple trigonometric series. Ann. Math., II. Ser. *66*, No. 2, 467–480, Zbl. 89,278.

Sjölin, P. (1971): Convergence almost everywhere of certain singular integrals. Ark. Mat. *9*, No. 1, 65–90, Zbl. 212,417.

Sobolev, S. L. (1950): Applications of functional analysis in mathematical physics. Leningrad: LGU (256 pp.). English translation: Providence, Rhode Island. Am. Math. Soc. 1963 (239 pp.) Zbl. 123,90.

Sokol-Sokolowski, K. (1947): On trigonometric series conjugate to Fourier series in two variables. Fundam. Math. *34*, 166–182, Zbl. 33,113.

Stein, E. M. (1958): Localization and summability of multiple Fourier series. Acta Math. *100*, No. 1–2, 93–147, Zbl. 85,284.

Stein, E. M. (1961): On certain exponential sums arising in multiple Fourier series. Ann. Math., II. Ser. *73*, No. 2, 87–109, Zbl. 99,55.

Stein, E. M. (1970): Singular integrals and differentiability properties of functions. Princeton: Princeton Univ. Press (290 pp.), Zbl. 207,135.

Stein, E. M. and Weiss, G. (1971): Introduction to Fourier analysis on Euclidean spaces. Princeton: Princeton Univ. Press, Zbl. 232.42007.

Sunouchi, Gen-Ichirô (1985): On the summability almost everywhere of multiple Fourier series at the critical index. Kodai Math. J. *8*, No. 1, 1–4, Zbl. 575.42017.

Szasz, O. and Minakshisundaram, S. (1947): On absolute convergence of multiple Fourier series. Trans. Am. Math. Soc. *61*, No. 1, 36–53, Zbl. 54,30.

Taylor, M. E. (1981): Pseudodifferential operators. Princeton: Princeton Univ. Press. (499 pp.), Zbl. 453.47026.

Tevzadze, N. R. (1970): On the convergence of the double Fourier series of a square summable function. Soobshch. Akad. Nauk Gruz. SSR *58*, No. 2, 277–279, [Russian] Zbl. 203,69.

Titchmarsh, E. C. (1958): Eigenfunction expansions associated with second order differential equations, II. Oxford: Clarendon Press (404 pp.), Zbl. 97,276.

Tonelli, L. (1928): Serie trigonometriche. Bologna: Nicole Zanichelli (527 pp.), Jbuch 54,298.

Trigub, R. M. (1980): Absolute convergence of Fourier integrals, summability of Fourier series and approximtion by polynomials of functions on the torus. Izv. Akad. Nauk SSSR, Ser. Mat. *44*, No. 6, 1378–1409. English translation: Math. USSR, Izv. *17*, 567–593 (1981), Zbl. 459.42019.

Wainger, S. (1965): Special trigonometric series in *k* dimensions. Mem. Am. Math. Soc. 59 (98 pp.), Zbl. 136,366.

Welland, G. V. and Ash, J. M. (1972): Convergence, uniqueness and summability of multiple Fourier series. Trans. Am. Math. Soc. *163*, 401–436, Zbl. 233.42014.

Yanushauskas, A. I. (1986): Multiple Fourier series. Novosibirsk: Nauka, Sibirskoe Otd. (272 pp.). [Russian] Zbl. 604.42012.

Yudin, V. A. (1979): A lower bound for the Lebesgue constant. Mat. Zametki *25*, No. 1, 119–122. English translation: Math. Notes *25*, 63–65 (1979), Zbl. 398.42016.

Zhizhiashvili, L. V. (1969): Conjugate functions and trigonometric series. Tbilisi: Izd-vo Tbilisk. Gos. Un-ta. [Russian]

Zhizhiashvili, L. V. (1970): On the divergence of Fourier series. Dokl. Akad. Nauk SSSR *194*, No. 4, 758–759. English translation: Sov. Math., Dokl. *11*, 1281–1283 (1971), Zbl. 217,140.

Zhizhiashvili, L. V. (1971): On convergence and divergence of Fourier series. Dokl. Akad. Nauk SSSR *199*, No. 6, 1234–1236. English translation: Sov. Math., Dokl. *12*, 1261–1263 (1971), Zbl. 259.42033.

Zhizhiashvili, L. V. (1973): On some questions in the theory of simple and multiple trigonometric series. Usp. Mat. Nauk *28*, No. 2, 65–119. English translation: Russ. Math. Surv. *28*, No. 2, 65–127 (1973), Zbl. 259.42027.

Zygmund, A. (1959): Trigonometric series I, II (2nd edition). Cambridge: Cambridge Univ. Press, Zbl. 85,56.

Zygmund, A. (1972): A Cantor-Lebesgue theorem for double trigonometric series. Stud. Math. *43*, No. 2, 173–178, Zbl. 214,325.

(1979) Harmonic analysis in Euclidean spaces. Proc. Symp. Pure Math. 35, part I, II. Providence, Rhode Island: Am. Math. Soc., Zbl. 407.00005, Zbl. 407.00006.

II. Methods of the Theory of Singular Integrals: Littlewood-Paley Theory and Its Applications

E. M. Dyn'kin

Translated from the Russian
by J. Peetre

Contents

Introduction

This article is an immediate continuation to the article "Methods of the Singular Integrals: Hilbert Transform and Calderón-Zygmund Theory", published in Vol. 15 of this series (Dyn'kin (1987)).

While Dyn'kin (1987) contains Chapters 1–3, this part contains Chapters 4 and 5. Notation and results from there will be freely employed below.

In September 1970 in the preface to his book (1970a) E. Stein wrote about the theory of singular integrals that "it (the subject) has an advanced degree of sophistication and is still rapidly developing, but has not yet reached the level of maturity that would require it to be enshrined in an edifice of great perfection". During the last two decades this rapid development has continued, and these words maintain their truth also today. Furthermore, we must now add to the short survey of results in the introduction to Dyn'kin (1987) some newer achievements, which were not mentioned there.

1. In the work of David, Journé, McIntosh, Meyer, Semmes the notion of weak boundedeess has been elucidated, and powerful and deep criteria for the L^2-boundedness of singular integrals have been established (the $T1$- and Tb-Theorems).

2. Coifman gave the first variant of Littlewood-Paley theory which does not use neither Fourier analysis nor properties of harmonic functions, and therefore extends to spaces of homogeneous type (for example, to Lipschitz surfaces in $\mathbb{R}^N$).

3. In the spring of 1987 Jones and Semmes found an elementary proof of the L^2-boundedness of the Cauchy integral on Lipschitz curves.

In the part at hand, we set forth, besides the classical result, also these developments obtained in the past ten years, mainly following preprints and summaries given at various meetings.

Calderón-Zygmund theory, as set forth in Chapter 3, shows how one can derive from the L^2 boundedness of a singular integral its boundedness in L^p, H^1, BMO, weighted spaces of vector functions etc. Its methods belong to the real variable theory and are geometrical, universal and easily adaptable. But Calderón-Zygmund theory is essentially a perturbation theory. It is applicable only if one knows an initial L^2 estimate. The main tool for obtaining L^2 estimates for singular integrals has in the past ten years been Littlewood-Paley theory.

What then is the Littlewood-Paley theory?

The following three typical examples give an idea what the subject which traditionally is referred to as *Littlewood-Paley theory* is about.

Example 1. The dyadic expansion. Let f be a summable function on the unit circle and denote its partial Fourier sums by $\{P_k(f)\}_0^\infty$. Set

$$g(f)(e^{i\theta}) = \left(\sum_{n=0}^{\infty} |P_{2^{n+1}}(f)(e^{i\theta}) - P_{2^n}(f)(e^{i\theta})|^2\right)^{1/2}.$$

It turns out that $f \in L^p(\mathbb{T})$, $1 < p < \infty$, if and only if $g(f) \in L^p(\mathbb{T})$. This is the classical *theorem of Littlewood-Paley on the dyadic expansion* (Stein (1970a), Zygmund (1959)). Recall further that our function f is given as the sum of the Bernshteĭn series

$$f = P_1 + \sum_{n=0}^{\infty} (P_{2^{n+1}} - P_{2^n}), \tag{1}$$

The Littlewood-Paley theorem expresses the membership of f in L^p in terms of the rate of convergence of this series.

Let now T be a Fourier series *multiplier*, that is, an operator acting on the Fourier coefficients via multiplication by a bounded sequence:

$$T : \hat{f}(k) \mapsto m(k)\hat{f}(k), \quad -\infty < k < \infty.$$

Then, in view of (1),

$$Tf = TP_1 + \sum_n (TP_{2^{n+1}} - TP_{2^n}).$$

In the term with index n there enter only values $m(k)$ with $2^n < |k| \le 2^{n+1}$ so, if $m(k)$ varies sufficiently slowly in this interval, it is not hard to obtain the estimate

$$g(Tf) \le Cg(f),$$

which shows that T is a bounded operator on all L^p spaces, $1 < p < \infty$. Thus from the Littlewood-Paley theorem one can obtain sufficient conditions for the boundedness of multipliers.

Example 2. Luzin function. Let f be a bounded analytic function in the upper halfplane $\mathbb{C}_+$. Define its *Luzin function* by the formula

$$S(f)(x) = \left(\iint_{|x-y|<t} |f'(y+it)|^2 dydt\right)^{1/2}, \quad x \in \mathbb{R}.$$

It turns out that $f \in H^p(\mathbb{C}_+)$, $0 < p < \infty$, if and only if $S(f) \in L^p(\mathbb{R})$. If $p > 1$ this is again a classical result (Stein (1970a), Zygmund (1959)). But that there is in this case too an integral representation analogous to (1) has been noticed only recently:

$$f(x) = -\frac{2}{\pi i} \iint_{\mathbb{C}_+} tf'(y+it) \frac{dydt}{(y-x-it)^2}, \quad x \in \mathbb{R}, \tag{2}$$

The condition $S(f) \in L^p$ controls the rate of convergence of this integral for $t \to 0$. The theorem that this implies that $f \in H^p$ for $0 < p \leq 1$ is due to Calderón (1965). The latter derived from this the following corollary: if $f, g \in H^2$ and $h' = f'g$ then $h \in H^1$. This result plays a key rôle in the work of Calderón (1965), (1977), (1980) on commutators of singular integrals and the Cauchy integral.

Example 3. Non-Poisson mean. Let ψ be a radial function in $\mathcal{D}(\mathbb{R}^n)$ such that $\int_{\mathbb{R}^n} \psi = 0$. For $f \in L^\infty(\mathbb{R}^n)$ set

$$Q_t f = \psi_t * f, \quad \psi_t(x) \stackrel{\text{def}}{=} t^{-n}\psi(x/t),$$

and

$$S(f)(x) = \left(\int_{|y-x|<t} |Q_t f(y)|^2 \frac{dydt}{t^{n+1}} \right)^{1/2}, \quad x \in \mathbb{R}^n.$$

In the last integral the integration is over the Luzin cone in $\mathbb{R}^{n+1}_+$ (Chapter 1, Sect. 1.1).

It turns out that $f \in L_p(\mathbb{R}^n)$, $1 < p < \infty$, if and only if $S(f) \in L_p(\mathbb{R}^n)$.

And again it turns out that not only this theorem of Littlewood-Paley type but also the integral representation

$$f = c_0^{-1} \int_0^\infty Q_t[Q_t f] \frac{dt}{t}, \tag{3}$$

c_0 being a constant depending on ψ, provide us with a very powerful tool. For example, let T be a singular integral with a Calderón-Zygmund kernel. Then formally

$$Q_t[Tf] = c_0^{-1} \int_0^\infty (Q_t T Q_s) Q_s f \frac{ds}{s}.$$

Estimating the operator T in $L_p(\mathbb{R}^n)$ then reduces to estimating $Q_t T Q_s$ in a corresponding function space in $\mathbb{R}^{n+1}_+$ to which $Q_s f$ belongs. But the kernel of $Q_t T Q_s$ is expressible in terms of the action of T on the the test function ψ_t and can be estimated explicitly. In this way one can prove such general boundedness criteria for T in L^2 as the $T1$-Theorem of David-Journé (cf. Sect. 1.2 of Chapter 5).

What is common in these three examples? In each case we have an integral representation where an auxiliary variable t or n appears, while the function f is represented as the image of a function in more variables: $P_{2^{n+1}} f(e^{i\theta}) - P_{2^n} f(e^{i\theta})$, $f'(y+it)$, $Q_t f(x)$. The condition $f \in L^p$ is expressed in the form of an estimate of the modulus of this new function, so that we get a parametric representation of L^p with a characterization of the class of the corresponding densities. This characterization reads as follows: one has to form a quadratic expression by first integrating with respect to the auxiliary variable (t or n) and then require that this quantity belongs to L^p in the original variables.

If one has a more or less explicit description of the operator T, then such a description leads to convenient criteria for L^p-boundedness.

Remark. It is possible to express in terms of the representations (1)–(3) not only L^p but also other function classes, notably classes of smooth functions such as Sobolev or Besov spaces. The description of Sobolev classes is parallel to the Littlewood-Paley theory (cf. Mazy'a (1985), Stein (1970a), Strichartz (1967), Triebel (1983)). The description of Besov spaces is simpler: one has first to take the L^p-norm in the original variables and then to take account of its behavior for $t \to 0$. For example, the Hölder class $\Lambda^\alpha(\mathbb{R}^n)$, $0 < \alpha < 1$, is described by the condition $\|Q_t f\|_{L^\infty(\mathbb{R}^n)} \le ct^\alpha$.

One has also variants of the Littlewood-Paley theory for analytic and harmonic functions, for semigroups, for martingales etc. (Durrett (1984), Folland and Stein (1982), Stein (1970b)). In Chapter 4 we describe some of this in some greater detail. In the seven first sections we discuss analytic functions in a halfplane and in Lipschitz domains, estimates for the distribution function, harmonic functions in $\mathbb{R}^n$, non-Poisson means, the Coifman construction and the dyadic expansion. In the short Sect. 8 we formulate a Littlewood-Paley type theorem for martingales and recent results by Meyer and collaborators on wavelets.

Chapter 5 is devoted to estimates of singular integrals in L^2 which follow from Littlewood-Paley theory.

Consider the *Hilbert transform* on the real axis:

$$Hf(x) = \text{P.V.}\frac{1}{\pi}\int_{-\infty}^{\infty} f(y)\frac{dy}{x-y}.$$

Let f and g be functions in $\mathcal{D}(\mathbb{R})$ with support in an interval $I \subset \mathbb{R}$. The estimate

$$|\langle Hf, g\rangle| \le c\|f\|_{L^2}\|g\|_{L^2}$$

is equivalent to the boundedness of H in L^2 and is hard to prove. It is however quite easy to obtain the equality

$$|\langle Hf, g\rangle| = \frac{1}{2}\left|\iint_{I\times I}[f(x)g(y) - f(y)g(x)]\frac{dxdy}{x-y}\right| \le c|I|^2\|f'\|_{L^2}\|g'\|_{L^2}. \quad (4)$$

Note however that (4) is not more rough than the L^2-estimate – because for "nice" functions with a regular behavior one has $\|f'\| \asymp |I|^{-1}\|f\|$, so that the estimate is not improvable as to "the order of magnitude", without replacing this norm by other norms. But in Example 3 we have seen that in the analysis of the operator $Q_t H Q_s$ it is precisely the action of H on "nice" scaling functions that matters. Moreover, there is the hope that in the scheme of Example 3 the estimate (4) replaces the L^2-estimate so that in this way one can establish the boundedness of H in L^2.

This hope turns out to materialize itself for the Hilbert transform and for any singular integral operator with a Calderón-Zygmund kernel. One can introduce a condition of "weak boundedness" for an operator, similar to (4), containing the norm of derivatives of f and g of any finite order with an appropriate degree of homogeneity in $|I|$. In particular, every operator with an antisymmetric kernel (cf. (4)) turns out to be weakly bounded. It turns out that if T is a weakly bounded operator such that $T1 = T^*1 = 0$, then it is bounded in L^2. The Hilbert transform is subject to this criterion. Quite generally, we know from Chapter 3 that for a Calderón-Zygmund operator $T1 \in \mathrm{BMO}$, because T acts from L^∞ into BMO. It turns out that there exist no other nontrivial boundedness conditions in L^2: if an operator with a Calderón-Zygmund kernel in $\mathbb{R}^n$ is weakly bounded and $T1 \in \mathrm{BMO}$, $T^*1 \in \mathrm{BMO}$, then it is bounded in L^2. This result is called the $T1$-*Theorem* and is due to David and Journé (1984). The $T1$-Theorem sheds light on a difference between translation invariant Calderón-Zygmund operators and general ones: for the former holds by necessity $T1 = T^*1 = 0$ and therefore translation invariant operator enjoy supplementary properties. For example, the condition $T1 = 0$ is sufficient for the boundedness of Calderón-Zygmund operators also in BMO and in Λ^α with $0 < \alpha < 1$.

The $T1$-Theorem became one of the main tools for obtaining L^2-estimates, especially after one had obtained a far reaching generalization of it, the *Tb-Theorem*, in which the rôle of the function 1 is taken over by a sufficiently general bounded function b (for example, subject to the condition $\operatorname{Re} b \geq \varepsilon > 0$).

In order to reduce the general case of the $T1$-Theorem to the case $T1 = T^*1 = 0$, one constructs from the operator T a special Calderón-Zygmund operator L_{ab} with $L_{ab}1 = a$, $L^*_{ab}1 = b$, where a and b are given functions in BMO. The construction of such an operator is a highly nontrivial matter. It is done with the help of integral representations of the Littlewood-Paley theory. The operator L_{ab} is called a *paraproduct*; its construction is described in Chapter 4.

Until recently the ultimate test for all L^2-boundedness criteria has been the question of L^2 boundedness of the Cauchy integral on a Lipschitz curve, that is, the boundedness in $L^2(\mathbb{R})$ of the operator

$$Tf(x) = \mathrm{P.V.} \int_{-\infty}^{\infty} \frac{1}{x - y + i[\varphi(x) - \varphi(y)]} f(y)\, dy, \tag{5}$$

where φ is a real valued absolutely continuous function with $\|\varphi'\|_{L^\infty} = M < \infty$. From an estimate of the operator (4) one can derive an estimate for general Calderón commutators (Sect. 2.4 of Chapter 3) with the kernel

$$\frac{1}{x-y} h\left\{\frac{\varphi(x) - \varphi(y)}{x - y}\right\}, \quad h \in C^\infty(\mathbb{R}),$$

along with their multidimensional analogues (Calderón (1977), (1980), Coifman, David, and Meyer (1983), Fabes, Jodeit, and Rivière (1978), Verchota

(1984)). Starting with the paper Calderón (1977), where Calderón established the boundedness of T for sufficiently small M, each new approach to L^2-estimates was tested on this example. By now one knows four different proofs of the boundedness of the operator (5).

1. The boundedness of T for all M was first proved by Coifman, McIntosh and Meyer (1982) employing a rather cumbersome method. The proof was later simplified in Coifman, Meyer, and Stein (1983).

2. The boundedness of T follows from the Tb-Theorem. Here the $T1$-Theorem is unsufficient, since it gives boundedness only for small values of M, as in Calderón (1977).

3. David (1984), (1986) and Murai (1983), (1984) developed a perturbation theory allowing to derive the boundedness of T for all M from the boundedness for small M.

4. Finally, in 1987 Jones and Semmes found an entirely elementary proof. In fact they established (with the aid of a simple geometric reasoning) that the Cauchy integral is bounded in $L^2(\Gamma)$ if for both domains, into which Γ cuts the plane, holds an analogue of the estimate $\|f\|_{H^2} \leq c\|S(f)\|_{L^2}$ in Example 2.

All four proofs are interesting and instructive, each opening up a new route. In Chapter 5 we set forth all these proofs, similarly to as we in Chapter 2 have discussed different approaches to the boundedness of the Hilbert transform. However, the Jones-Semmes proof will be given in the very beginning of the part, in Sect. 2 of Chapter 4, directly after our discussion of the necessary Littlewood-Paley estimates. Besides its simplicity, it is remarkable from the point of view that it gives back to the problem of the Cauchy integral its rôle as an interesting but particular example in Littlewood-Paley theory, further removing some awkward points of the theory.

Besides weak boundedness and the Cauchy integral on Lipschitz curves, we discuss in Chapter 5 in detail David's theorem on the Cauchy integral on Carleson curves, to which we also return in Sect. 6 of Chapter 2.

Finally, the last section of Chapter 5 collects some separate developments on singular integral operators, which for various reasons were not considered in Chapter 2 and 3. This comprises a new proof of the boundedness of the Hilbert transform (Meyer (1985), Wittman (1987)) and some references concerning the boundedness of singular integrals in spaces of smooth functions and in BMO.

The enumeration of the formulae and the theorems is continuous in each Chapter.

The Jones-Semmes proof of the boundedness of the Cauchy integral was kindly communicated to the author by Paul Koosis. Discussions with S. V. Kislyakov have had a decisive influence on the text of this article and on the organization of the material, much of which again was made available to us by Yves Meyer. To all three mathematicians I hereby express my sincere thanks.

Chapter 4
Littlewood-Paley Theory

A general introduction to Littlewood-Paley theory is given in the introduction of this part; cf. further Coifman and Weiss (1978), Cowling (1981), Stein (1970a), (1982), Zygmund (1959).

In Sect. 1 we describe a variant of the theory connected with analytic functions and the Luzin area integral. We begin with the definition of the Luzin function and conclude with Calderón's commutator theorem (1965). In this way we encounter with all the main ideas of the theory, so that Sect. 1 serves also as an introduction to the entire chapter, where these ideas will be developed in a more general setting.

In Sect. 2 an analogue of Littlewood-Paley theory will be developed for analytic functions in plane Lipschitz domains. A central result here is the proof of the L^2-boundedness of the Cauchy integral on Lipschitz curves due to Jones and Semmes. As a preparation of this proof we develop an elementary approach based on conformal mapping, leaving more powerful methods founded on harmonic estimates to a following section.

In Sect. 3 we set forth the Burkholder-Gundy theorem on estimating the distribution function which allows one to connect the Luzin function with non-tangential maximal functions (cf. Sect. 6.3 of Chapter 3). Dahlberg extended this theorem to harmonic functions in multidimensional Lipschitz domains. Dahlberg's proof is difficult so we illustrate his idea at the hand of a simple example isolating the technical difficulties in estimating a harmonic measure.

Sections 1–3 give a closed presentation of the "analytic" Littlewood-Paley theory. In the remaining sections of the chapter various extensions of the theory to other objects are described.

In Sect. 4 harmonic functions in $\mathbb{R}^n$ and in Lipschitz domains are considered. This theory is quite close to the "analytic" theory (more exactly, the most recent proofs in the theory of analytic functions are specially constructed with applications also to the harmonic case in mind). Therefore, the contents of Sect. 3 have an expository character; we indicate the changes necessary in the formulations of Sects. 1–3.

In Sect. 5 non-Poisson means are described in the case when the solution of Dirichlet's problem is replaced by the convolution of the given function with an arbitrary smooth kernel (very often a finite[1] one). This variant of Littlewood-Paley theory has noteworthy implications for the study of singular integrals; in Chapter 5 we will use it frequently. In Sect. 5.4 we introduce the paraproducts, about which we spoke already in the introduction.

The classical Littlewood-Paley theory is not sufficiently flexible, because the key L^2-estimates are based either on the Fourier transform or on Green's

[1] *Translator's note.* That is, with compact support (cf. p. 34).

formula. Recently Coifman has given a variant of the theory which is free from this incoveniency and is applicable to any Lipschitz manifold. Coifman's construction is, following David, Journé, and Semmes (1986), outlined in Sect. 6.

Section 7 is devoted to the dyadic expansion and to the Marcinkiewicz-Mikhlin-Hörmander multiplier theorem. In this section our presentation mainly follows Stein's book [83].

The short Sect. 8 is devoted to two variants of the theory, which, because of lack of space, cannot be described in detail here. It is a question of Littlewood-Paley type inequalities for martingales and a recent result by Meyer and collaborators on wavelet expansions.

§1. The Luzin Function

1.1. Definition and Simplest Properties. Let f be an analytic function in the upper halfplane $\mathbb{C}_+$, assuming for simplicity that $f \in H^p$, where $1 \le p < \infty$.

The *area integral* or the *Luzin function* $S(f)$ is defined by the formula

$$S_\alpha(f)(x) = \left(\iint_{\Gamma_\alpha(x)} |f'(\zeta)|^2 \, d\xi d\eta \right)^{1/2}, \quad x \in \mathbb{R}.$$

Here $\zeta = \xi + i\eta$ and

$$\Gamma_\alpha(x) = \{\zeta : |\xi - x| < \alpha\eta\}$$

is the *Luzin sector.* Usually the size of the opening α of the sector does not play any rôle and then we can omit the α in the notation. It is clear that $S(f)^2(x)$ is the area of the image of $\Gamma(x)$ under the (polyvalent) conformal map f. This is why $S(f)$ is called the area integral.

The function $S(f)$ controls the convergence of $f(\zeta)$ to $f(x)$ as ζ approaches x through $\Gamma(x)$, that is, nontangential convergence. For example, if $S(f)(x) < \infty$ for $x \in E \subset \mathbb{R}$, then $f(\zeta)$ tends a. e. on E to $f(x)$ (Carleson (1967), Garnett (1981)).

It is clear that $S_\alpha \le S_\beta$ if $\alpha \le \beta$. It follows readily from the mean value theorem for analytic functions that

$$D_\beta(f)(x) \stackrel{\text{def}}{=} \sup_{\Gamma_\beta(x)} |f'(\zeta)| \cdot \eta \le c_{\alpha,\beta} \, S_\alpha(f)(x) \tag{4.1}$$

for $\beta < \alpha$. On the other hand, by the mean value theorem

$$D_\beta(f)(x) \le c_{\alpha,\beta} \, f^*_\alpha(x), \quad \beta < \alpha, \tag{4.2}$$

where f^*_α is the *nontangential maximal function* (cf. Chapter 1, Sect. 7.2).

The Luzin function $S(f)$ can likewise be defined for analytic functions in the disc $\mathbb{D}$. In this case

$$S_\alpha(f)(e^{i\theta}) = \left(\iint_{\Gamma_\alpha(\theta)} |f'(\zeta)|^2 \, d\xi d\eta \right)^{1/2},$$

where $\Gamma_\alpha(\theta) = \{\zeta \in \mathbb{D} : |\zeta - e^{i\theta}| \le (1+\alpha)(1-|\zeta|)\}$.

In the next section we will encounter analogues of $S(f)$ for functions analytic in an arbitrary Lipschitz domain in the plane.

1.2. L^2-Estimates. It is clear that

$$\iint_{\mathbb{C}_+} S_\alpha(f)^2(x)\, dx = c_\alpha \iint_{\mathbb{C}_+} y|f'(x+iy)|^2 \, dxdy,$$

where the coefficient c_α depends only on α.

Theorem 4.1.

$$\iint_{\mathbb{C}_+} y|f'(x+iy)|^2 \, dxdy = \frac{1}{4} \iint_{\mathbb{C}_+} |f(x)|^2 \, dx.$$

Corollary. $\|S(\alpha)(f)\|_{L^2} \asymp \|f\|_{L^2}$ *for* $\alpha > 0$.

Theorem 4.1 can be proved in several ways.

(i) It follows at once upon applying Green's formula (Sect. 1 of Chapter I) to the domain $\mathbb{C}_+$ and the functions $|f|^2$ and y. It is easy to overcome the formal complications (that $|f|^2$ is not smooth, that $\mathbb{C}_+$ is unbounded). Note the twofold rôle of y in this reasoning: it is a harmonic function in $\mathbb{C}_+$ and at the same time the distance to the boundary.

(ii) $f_y(x) \overset{\text{def}}{=} f(x+iy)$ is the Poisson integral of $f(x)$ (Sect. 7.1 of Chapter 1). Taking the Fourier transform with respect to x gives $\hat{f}_y(t) = e^{-y|t|}\hat{f}(t)$ so that Theorem 4.1 follows from Plancherel's formula.

The analogue of Theorem 4.1 for the disk reads:

$$\iint_{\mathbf{D}} |f'(z)|^2 \log \frac{1}{|z|} \, dxdy = \frac{1}{4} \int_{-\pi}^{\pi} |f(e^{i\theta}) - f(0)|^2 \, d\theta. \tag{4.3}$$

From this we see that the two-sided estimate $\|S(f)\|_{L^2} \asymp \|f\|_{L^2}$ in the disk requires the additional normalization $f(0) = 0$ (or $f(z_0) = 0$ where z_0 is a fixed point of $\mathbb{D}$). In the halfspace the normalization is implicitly contained in the condition $f \in H^p$, $1 \le p < \infty$, as then $\lim\limits_{y\to\infty} f(x+iy) = 0$.

A detailed discussion of (4.3) can be found in the book Garnett (1981). In general the whole Littlewood-Paley theory can be literally transferred to the disc, where one instead of Fourier integrals studies Fourier series. We will in general not state explicitly the "periodic" analogues of our theorems.

1.3. L^p-Estimates. The standard technique of Chapter 3 (Calderón-Zygmund operators) allows one to derive from Theorem 4.1 L^p-estimates for $1 < p < \infty$. To this end let us introduce the Hilbert space H of functions φ in the sector $\Gamma = \Gamma(0)$ with the norm

$$\|\varphi\|_H = \left(\iint_\Gamma |\varphi(\zeta)|^2 \, d\xi d\eta\right)^{1/2}.$$

Define $K(x)$ for each $x \in \mathbb{R}$ by the formula

$$[K(x)](\zeta) = \frac{1}{2\pi i} \frac{1}{(\zeta + x)^2}, \qquad \zeta \in \Gamma,$$

so that $K(x)$ is an element of H. Consider the integral operator

$$Tf(x) = \int_{-\infty}^{\infty} K(x-y) f(y) \, dy,$$

acting from $L^2(\mathbb{R})$ into the space of H-valued vector functions $L^2(\mathbb{R}, H)$. If $f \in H^2(\mathbb{C}_+)$ then clearly $[Tf(x)](\zeta) = f'(x+\zeta)$ so that $\|Tf(x)\|_H = S(f)(x)$ and, in view of Theorem 4.1, $\|Tf\|_{L^2(\mathbb{R},H)} = c_\alpha \|f\|_{L^2}$. If $f \in L^2 \ominus H^2$ then by Cauchy's theorem $Tf = 0$. Hence, T is bounded in L^2. But simple estimates show that K is a Calderón-Zygmund kernel, so that T is a Calderón-Zygmund operator. In view of the results of Chapter 3, T is bounded in L^p for all p, $1 < p < \infty$, and further for all weighted spaces L^p_w with $w \in (\mathrm{A}_p)$. As T is isometric on H^2, a duality argument at once gives also the inverse estimate. We have arrived at the following theorem.

Theorem 4.2. *If $1 < p < \infty$ and $w \in (A_p)$ then*

$$\|S_\alpha(f)\|_{L^p_w} \asymp \|f\|_{L^p_w}, \qquad \alpha > 0,$$

for all functions $f \in H^p_w$.

Let now $f \in H^1$. Then f can be written as the product of two H^2-functions (cf. Garnett (1981), Koosis (1980)):

$$f = gh, \quad \|g\|_{H^2} = \|h\|_{H^2} = \|f\|_{H^1}^{1/2}.$$

It follows that for all $\zeta \in \Gamma(x)$

$$|f'(\zeta)| = |g'(\zeta)h(\zeta) + g(\zeta)h'(\zeta)| \le |g'(\zeta)|\, h^*(x) + g^*(x)\, |h'(\zeta)|,$$

whence

$$S(f)(x) \le h^*(x) S(g)(x) + g^*(x) S(h)(x),$$
$$\|S(f)\|_{L^1} \le \|h^*\|_{L^2} \|S(g)\|_{L^2} + \|g^*\|_{L^2} \|S(h)\|_{L^2} \le c\|g\|_{L^2}\|h\|_{L^2} = c\|f\|_{H^1}$$

(recall that $\|g^*\|_{L^2} \le c\|g\|_{L^2}$ by the maximal theorem). Thus, the estimate

$$\|S(f)\|_{L^p} \le c_p\|f\|_{H^p} \tag{4.4}$$

remains in force also for $p = 1$. In the same way one proves it for $0 < p < 1$ also.

To estimate $S(f)$ from below is a different matter. It turns out that the inverse estimate

$$\|f\|_{H^p} \le C_p\|S(f)\|_{L^p} \tag{4.5}$$

for analytic functions in a halfspace holds true also for $0 < p \le 1$, but it is much harder to prove this result. For $p = 1$ this estimate was first obtained in Calderón (1965). In order to understand the origin of (4.5), we require new methods.

1.4. Integral Representation. Let again f be analytic in the upper halfplane $\mathbb{C}_+$ and assume, for instance, that $f \in H^p$, $1 \le p < \infty$. Then

$$f(x) = -4\int_0^\infty \eta f''(x+2i\eta)d\eta = -\frac{2}{\pi i}\int_{-\infty}^\infty \eta d\eta \int_0^\infty \frac{f'(\xi+i\eta)}{(\xi-x-i\eta)^2}\,d\xi.$$

In this identity we have expressed $f''(x+2i\eta)$ in terms of the Cauchy integral of $f'(\xi+i\eta)$ along the horizontal line $\{\xi+i\eta, -\infty < \xi < \infty\}$. Hence

$$f(x) = -\frac{2}{\pi i}\iint_{\mathbb{C}_+} f'(\xi+i\eta)\frac{\eta}{(\xi-x-i\eta)^2}\,d\xi d\eta. \tag{4.6}$$

In other words, the analytic function f is represented as a potential

$$f(x) = \iint_{\mathbb{C}_+} \frac{\varphi(\xi,\eta)}{(\xi-x-i\eta)^2}\,d\xi d\eta, \quad x \in \mathbb{R}, \tag{4.7}$$

where

$$\varphi(\xi,\eta) = -\frac{2}{\pi i}\eta f'(\xi+i\eta).$$

Formula (4.7) allows us to pass from the direct estimates for $S(f)$ to the inverse ones and provides a universal tool for the entire theory. Note that the density φ in (4.7) is not defined uniquely – the integral representation is quite "free".

Let us introduce the analogue of *Luzin's function* for the potential (4.7):

$$A_\alpha(\varphi)(x) = \left(\iint_{\Gamma_\alpha(x)} |\varphi(\xi,\eta)|^2\,\frac{d\xi d\eta}{\eta^2}\right)^{1/2}.$$

Theorem 4.3. *The analytic function f is in H^p, $1 < p < \infty$, if and only if it can be represented as a potential* (4.7) *with $A_\alpha(\varphi) \in L^p(\mathbb{R}^n)$.*

Indeed, if $f \in H^p$ then it admits the representation (4.7) with

$$A(\varphi) = S(f) \in L^p.$$

Conversely, assume that we have the representation (4.7) with $A(\varphi) \in L^p$. Then it is clear that f has an analytic continuation to $\mathbb{C}_+$, so it suffices to show that

$$\left|\int_{-\infty}^{\infty} f(x)g(x)\,dx\right| \le c\|g\|_{L^{p'}}$$

for every $g \in H_-^{p'}$ (Garnett (1981)). ($H_-^{p'}$ is the space $H^{p'}$ for the lower halfplane $\mathbb{C}_-$, $p' = \frac{p}{p-1}$ being the conjugate exponent). However

$$\begin{aligned}\int_{-\infty}^{\infty} f(x)g(x)\,dx &= \iint_{\mathbb{C}_+} \varphi(\xi,\eta)\,d\xi d\eta \int_{-\infty}^{\infty} \frac{g(x)}{(\xi - i\eta - x)^2}\,dx \\ &= 2\pi i \iint_{\mathbb{C}_+} \varphi(\xi,\eta) g'(\xi - i\eta)\,d\xi d\eta. \end{aligned} \tag{4.8}$$

But for any function $F(\xi,\eta)$ Fubini's theorem yields:

$$\iint_{\mathbb{C}_+} F(\xi,\eta)\,d\xi d\eta = c_\alpha \int_{-\infty}^{\infty} dx \iint_{\Gamma_\alpha(x)} F(\xi,\eta)\,\frac{d\xi d\eta}{\eta}, \tag{4.9}$$

where c_α depends on α only. Hence

$$\begin{aligned}\left|\iint_{\mathbb{C}_+} \varphi(\xi,\eta) g'(\xi - i\eta)\,d\xi d\eta\right| &= c_\alpha \left|\int_{-\infty}^{\infty} dx \iint_{\Gamma_\alpha(x)} \varphi \cdot g'\,\frac{d\xi d\eta}{\eta}\right| \\ &\le c_\alpha \int_{-\infty}^{\infty} dx \left(\iint_{\Gamma_\alpha(x)} |\varphi|^2 \frac{1}{\eta^2}\right)^{1/2} \left(\iint_{\Gamma_\alpha(x)} |g'|^2\right)^{1/2} \\ &= c_\alpha \int_{\infty}^{\infty} A(\varphi)(x) S(g)(x)\,dx. \end{aligned}$$

As

$$\int_{\infty}^{\infty} A(\varphi)S(g) \le \|A(\varphi)\|_{L^p} \|S(g)\|_{L^{p'}} \le c\|A(\varphi)\|_{L^p} \|g\|_{L^{p'}},$$

we conclude that $f \in H^p$.

Is Theorem 4.3 true for $p = 1$? If $f \in H^1$ we have, in view of (4.4), the imbedding $A(\varphi) = S(f) \in L^1$. The converse is harder.

1.5. H^1-Estimates

Theorem 4.4. *If the function f admits the representation* (4.77) *with $A(\varphi) \in L^1(\mathbb{R})$ then $f \in H^1$ and $\|f\|_{H^1} \le c\|A(\varphi)\|_{H^1}$.*

Corollary (Calderón (1965)). *If* $S(f) \in L^1$ *then* $f \in H^1$ *and* $\|f\|_{H^1} \le c\|S(f)\|_{L^1}$.

Theorem 4.4 was proved by Coifman, Meyer and Stein in two recent papers (1983), (1985), using very promising ideas due to Chang and Fefferman (1980), (1985).

Let us consider the space $\mathcal{F}$ of all functions φ in $\mathbb{C}_+$ with finite norm $\|\varphi\|_{\mathcal{F}} = \|A(\varphi)\|_{L^1}$. A function $a \in \mathcal{F}$ is termed an *atom* if $\operatorname{supp} a \subset \square(I)$ and

$$\iint_{\square(I)} |a(\xi,\eta)|^2 \frac{d\xi d\eta}{\eta} \le \frac{1}{|I|}.$$

It is easy to see that

$$\|a\|_{\mathcal{F}} = \|A(a)\|_{L^1} \le c|I|^{1/2}\|A(a)\|_{L^2} \le \text{const}.$$

Lemma 4.5. *If* $a \in \mathcal{F}$ *is an atom and*

$$h(x) = \iint_{\mathbb{C}_+} a(\xi,\eta)\frac{d\xi d\eta}{(x-\xi+i\eta)^2}, \quad x \in \mathbb{R},$$

then $h \in H^1$ *and* $\|h\|_{H^1} \le C < \infty$.

Indeed, in view of Theorem 4.3

$$\|h\|_{L^2} \le c\|A(a)\|_{L^2} \le c|I|^{-1/2}.$$

This means that the contribution of the interval I to $\|h\|_{L^1}$ is, in view of Schwarz's inequality, bounded. The contribution of $\mathbb{R}\backslash 2I$ is easily estimated directly.

Now Theorem 4.4 follows from the following general result.

Theorem 4.6. *Every function* $\varphi \in \mathcal{F}$ *can be written in the form*

$$\varphi = \sum_{k=1}^{\infty} \lambda_k a_k,$$

where the a_k *are atoms and* $\sum_1^\infty |\lambda_k| \le c\|\varphi\|_{\mathcal{F}}$.

The proof of Theorem 4.6 is immediate; for details we refer to Coifman, Meyer, and Stein (1983), (1985). Here we just indicate the construction of the atoms. Set $E_k = \{x : A(\varphi)(x) > 2^k\}$ and $\Omega_k = \{x : M\chi_{E_k}(x) > 1/2\}$, $-\infty < k < \infty$. The open set Ω_k is the union of intervals I_{kj}. Set

$$\triangle_{kj} = \square(I_{kj})\backslash \left\{\bigcup_l \square(I_{kj} \cap I_{k+1,l})\right\}.$$

Then the decomposition $\varphi = \sum \varphi\chi_{\Delta_{kj}}$ is the sought one, and all terms are atoms up to normalization of the coefficent λ_{kj}.

We record the following consequence of the estimate $\|f\|_{H^1} \le c\|S(f)\|_{L^2}$:

Theorem 4.7. *If $f, g \in H^2$, $h \in H^p$ for some $p > 0$ and $h' = f'g$, then $h \in H^1$ and $\|h\|_{H^1} \le c\|f\|_{H^1}\|g\|_{H^1}$.*

Indeed, in the hypothesis of Theorem 4.7 $S(h) \le g^* S(f)$.

1.6. Application: The Commutator Theorem. In Calderón (1965), where this author for the first time proved Theorem 4.7, Calderón also introduced a new class of singular integral operators, which in Sect. 2.4 of Chapter 3 were called *Calderón commutators.* These are operators of the form

$$Tf(x) = \text{P.V.} \int_{-\infty}^{\infty} \frac{\varphi(x) - \varphi(y)}{(x-y)^2} f(y)\, dy,$$

where φ ia a Lipschitz function on $\mathbb{R}$, that is, $\varphi' \in L^\infty$. In Sect. 2.4 of Chapter 3 we have seen that the kernel of such an operator is a regular Calderón-Zygmund kernel.

Theorem 4.8. *The operator T is bounded on $L^2(\mathbb{R})$ and, thus, is a Calderón-Zygmund operator.*

We may assume that $f, \varphi \in \mathcal{D}(\mathbb{R})$ and consider instead of the principal value the regularization

$$Qf(x) = \lim_{\delta \to +0} \int_{-\infty}^{\infty} \frac{\varphi(x) - \varphi(y)}{(x - y + i\delta)^2} f(y)\, dy.$$

Then it easy to see that $|Tf - Qf| \le cMf$, so that it suffices to establish the boundedness of Q. Set $e(t) = \chi_{(0,\infty)}(t)$; write $\varphi(x) - \varphi(y)$ in the form

$$\varphi(x) - \varphi(y) = \int_{-\infty}^{\infty} \varphi'(s)[e(x-s) - e(y-s)]\, ds.$$

In view of Sect. 2.2 of Chapter 3 we have the representation $f = f_+ + f_-$, with $f_\pm \in H^2(\mathbb{C}_\pm)$. Finally, let $g \in \mathcal{D}(\mathbb{R})$, which we likewise write as $g = g_+ + g_-$. Then

$$\begin{aligned}
&\int_{-\infty}^{\infty} Qf(x)g(x)\, dx \\
&\quad = \lim_{\delta \to +0} \int_{-\infty}^{\infty} \varphi'(s)\, ds \int_s^{\infty} \{g(x)f'_+(x+i\delta) + f(x)g'_-(x-i\delta)\} dx \\
&\quad = \int_{-\infty}^{\infty} \varphi'(s)\, ds \left\{ f_+(s)g_-(s) + \int_s^{\infty} [g_+(x)f'_+(x) + f_-(x)g'_-(x)] dx \right\}.
\end{aligned}$$

But $\|f_\pm\|_{H^2} \le \|f\|_{L^2}$, $\|g_\pm\|_{H^2} \le \|g\|_{L^2}$, so that by Theorem 4.7 the function within the brackets is summable and its L^1-norm does not exceed the product $\|f\|_{L^2}\,\|g\|_{L^2}$. Hence

$$\left|\int_{-\infty}^{\infty} Qf \cdot g\right| \le c\|\varphi'\|_{L^\infty}\|f\|_{L^2}\|g\|_{L^2},$$

so that the operator Q – and thereby also T – is bounded in L^2.

Theorem 4.8 was the starting point for numerous papers on estimates of commutators and became also the model for Calderón's famous paper (1977) on the Cauchy integral.

1.7. BMO-Estimates. The analogue of the previous theorem for $p = \infty$ is, of course, the estimate of the BMO-norm of the function f in terms of f'. But this estimate cannot any longer be given in terms of the area integral. Recall (Sect. 2.3 of Chapter 1) that the measure μ in $\mathbb{C}_+$ is said to be a *Carleson measure* if

$$\kappa(\mu) = \sup_I \mu(\Box(I))/|I| < \infty,$$

where the supremum is carried over all intervals $I \subset \mathbb{R}$.

Theorem 4.9. *An analytic function f on $\mathbb{C}_+$ belongs to* BMO($\mathbb{R}$) *if and only if the measure μ_f,*

$$d\mu_f = \eta|f'(\xi + i\eta)|^2\, d\xi d\eta,$$

is a Carleson measure. Moreover, we have

$$\kappa(\mu_f) \asymp \|f\|^2_{\mathrm{BMO}}.$$

The proof of Theorem 4.9 can be found in Fefferman and Stein (1972), Garnett (1981). In one direction the proof is easy: if $f \in$ BMO and $I \subset \mathbb{R}$ is an interval, then

$$f = f_{2I} + (f - f_{2I})\chi_{2I} + (f - f_{2I})\chi_{\mathbf{R}\setminus 2I} = f_{2I} + f_1 + f_2.$$

Here f_{2I} is the average of f over the interval $2I$. A constant term has no influence on $\mu_f(\Box(I))$. Furthermore,

$$\mu_{f_2}(\Box(I)) \le \iint_{\mathbb{C}_+} \eta|f_1'|^2\, d\xi d\eta = \frac{1}{4}\int_{-\infty}^{\infty} |f_1|^2 \le |I|\,\|f\|^2_{\mathrm{BMO}}.$$

Finally, if $\zeta \in \Box(I)$, then, in view of Cauchy's formula,

$$|f_2'(\zeta)| \le \iint_{|y-x_0|<|I|} \frac{|f(y) - f_{2I}|}{|y - x_0|^2}\, dy \le \frac{C}{|I|}\|f\|_{\mathrm{BMO}},$$

in view of Sect. 9.1 of Chapter 1, so that $\mu_{f_2}(\Box(I)) \le c\|f\|^2_{\text{BMO}}|I|$.

The converse is much deeper and equivalent to Fefferman's theorem on the duality of H^1 and BMO. Assume that $\kappa(\mu_f) < \infty$. It follows from the integral representation (4.6) that f can be written as a potential (4.7) with

$$B(\varphi)(x) = \sup_{I \ni x} \left(\frac{1}{|I|} \iint_{\Box(I)} |\varphi(\xi,\eta)|^2 \frac{d\xi d\eta}{\eta} \right)^{1/2} \le \kappa(\mu_f)^{1/2} < \infty.$$

In view of Fefferman's theorem it suffices to verify that

$$\left| \int_{-\infty}^{\infty} f(x) g(x)\, dx \right| \le c\|B(\varphi)\|_{L^\infty} \|g\|_{L^1}$$

for every function $g \in H^1(\mathbb{C}_-)$. But, in view of (4.8), this estimate follows from the following lemma (Coifman, Meyer, and Stein (1983), (1985), cf. Fefferman and Stein (1972)).

Lemma 4.10. *For arbitrary functions φ and ψ on $\mathbb{C}_+$ the following holds:*

$$\iint_{\mathbb{C}_+} |\varphi(\xi,\eta)\psi(\xi,\eta)| \frac{d\xi d\eta}{\eta} \le c \int_{-\infty}^{\infty} B(\varphi)(x)\, A(\psi)(x)\, dx.$$

As $A(\psi) = S(g) \in L^1$ for $\psi = \eta g'$, this implies Theorem 4.9.

Set $\Gamma^\varepsilon(x) = \{\zeta \in \Gamma(x), 0 < \eta < \varepsilon\}$ and

$$A(\varphi,\varepsilon)(x) = \left(\int_{\Gamma^\varepsilon(x)} |\varphi(\xi,\eta)|^2 \frac{d\xi d\eta}{\eta^2} \right)^{1/2} \le A(\varphi)(x).$$

Define $\varepsilon(x)$ by the formula

$$\varepsilon(x) = \sup\{\varepsilon : A(\varphi,\varepsilon)(x) \le MB(\varphi)(x)\},$$

where M is a sufficiently large constant (which will be determined later). Then we have for any interval $I \subset \mathbb{R}$

$$\text{mes}\{x \in I : \varepsilon(x) \ge |I|\} \ge c_M |I|, \tag{4.10}$$

with c_M depending on M only. Indeed, by Fubini's theorem,

$$\frac{1}{|I|} \int_I A(\varphi, |I|)^2(x)\, dx \le c \frac{1}{|I|} \int_{\Box(3I)} |\varphi|^2 \frac{d\xi d\eta}{\eta} \le c \inf_{x \in I} B(\varphi)^2(x),$$

which gives (4.10) provided $M^2 > c$. Next, by Fubini's theorem and Schwarz's inequality, we deduce

$$\begin{aligned}
&\iint_{\mathbb{C}_+} |\varphi| \cdot |\psi| \frac{d\xi d\eta}{\eta} \\
&\quad \le c_M^{-1} \int_{-\infty}^{\infty} dx \iint_{\Gamma^{\varepsilon(x)}(x)} |\varphi| \cdot |\psi| \frac{d\xi d\eta}{\eta^2} \\
&\quad \le c_M^{-1} \int_{-\infty}^{\infty} dx \left(\iint_{\Gamma^{\varepsilon(x)}(x)} |\varphi|^2 \frac{d\xi d\eta}{\eta^2} \right)^{1/2} \left(\iint_{\Gamma^{\varepsilon(x)}(x)} |\psi|^2 \frac{d\xi d\eta}{\eta^2} \right)^{1/2} \\
&\quad = c_M^{-1} \int_{-\infty}^{\infty} A(\varphi, \varepsilon(x))(x)\, A(\psi, \varepsilon(x))(x)\, dx \\
&\quad \le M c_M^{-1} \int_{-\infty}^{\infty} B(\varphi)(x)\, A(\psi)(x)\, dx,
\end{aligned}$$

establishing the lemma.

Corollary. $f \in \mathrm{BMO}$ *if and only if it can be represented as a potential (4.7) with* $B(\varphi) \in L^2$, *and then*

$$\|f\|_{\mathrm{BMO}} \asymp \|B(\varphi)\|_{L^\infty}.$$

Remark. It is well-known that for $f \in \mathrm{BMO}$ we have $|f'(\xi + i\eta)| \le c\|f\|_{\mathrm{BMO}}\, \eta^{-1}$. Therefore the measure $|f'(\xi + i\eta)| d\xi d\eta$ majorizes μ_f. But it need not be a Carleson measure. However, it turns out that (Garnett (1981)) for $f \in \mathrm{BMO}$ one can always find a function F in $\mathbb{C}_+$ (not analytic and not even harmonic) such that

(i) $F \in C^\infty(\mathbb{C}_+)$,
(ii) $\sup_{y>0} |F(x,y)| \in L^1_{\mathrm{loc}}(\mathbb{R})$,
(iii) $\lim_{y\to 0} F(x,y) = f(x)$ a. e. and
(iv) $|\nabla F(x,y)| dx dy$ is a Carleson measure.

The correspondence $f \mapsto F$ is non-linear.

§2. Lipschitz Domains and Cauchy Integral

In Sect. 1 we have constructed the Littlewood-Paley theory for analytic functions in a halfplane. Now we turn to functions analytic in Lipschitz domains.

2.1. Conformal Maps of Lipschitz Domains. The definition of Lipschitz domains in $\mathbb{R}^n$ was discussed already in Sect. 1.2 of Chapter 1. Here we will restrict ourselves to the important special case of *special Lipschitz domains* of the form

$$G = \{(x, y) \in \mathbb{R}^2 : y > \varphi(x)\},$$

where f is a Lipschitz function on $\mathbb{R}$, that is,

$$|\varphi(x) - \varphi(y)| \leq M|x - y|, \quad x, y \in \mathbb{R},$$

Any Lipschitz domain in $\mathbb{R}^2$ is locally of this form. The least number $M = \|\varphi'\|_{L^\infty}$ is called the *Lipschitz constant* of G. Vertical shifts $(x, y) \mapsto (x, y+t)$, $t \geq 0$, map G into itself. In particular, the boundary

$$\partial G = \{(x, y) : y = \varphi(x)\}$$

gives rise to a family of parallel lines

$$\Lambda_t = \{(x, y) : y = \varphi(x) + t\}, \quad t > 0.$$

Consider a conformal map ψ of the upper halfplane $\mathbb{C}_+$ onto G such that $\psi(\infty) = \infty$ and $\operatorname{Re}\psi' \geq 0$. It is well-known (Goluzin (1966), Privalov (1959)) that the function ψ' is outer in the sense of Beurling in $\mathbb{C}_+$ and that

$$|\arg \psi'| \leq \arctan M < \frac{\pi}{2}.$$

In particular

$$|\psi'| \leq c_M \cdot \operatorname{Re}\psi', \quad c_M = \sqrt{M^2 + 1}. \tag{4.11}$$

Consequently, the weight $w = |\psi'|$ on $\mathbb{R}$ satisfies the Muckenhoupt condition (A_2) (cf. Sect. 2.3 of Chapter 2; besides, this can also be verified easily directly).

The measure μ_G,

$$d\mu_G = y \left| \frac{\psi''(z)}{\psi'(z)} \right|^2 dxdy, \quad z = x + iy,$$

is a Carleson measure on $\mathbb{C}_+$. Indeed, $|\frac{\psi''}{\psi'}|^2 = |\frac{d}{dz} \log \psi'|^2$ and $\log|\psi'| \in \operatorname{BMO}(\mathbb{R})$ so the fact that μ_G is a Carleson measure follows from Sect. 1.7.

Let us consider in G the *Luzin sector* ($\alpha > 0$)

$$\Gamma_\alpha(z) = \{\zeta \in G : |\zeta - z| < (1 + \alpha)\rho(\zeta, \partial G)\}, \quad z \in \partial G.$$

By the distortion theorems of Koebe and Lavrent'ev (cf. Belinskiĭ (1974), Goluzin (1966)) we can find constants α_1 and α_2 such that for $z = \psi(x)$, $x \in \mathbb{R}$,

$$\psi[\Gamma_{\alpha_1}(x)] \subset \Gamma_\alpha(z) \subset \psi[\Gamma_{\alpha_2}(x)], \tag{4.12}$$

Finally, the Koebe distortion theorem gives:

$$\rho[\psi(z), \partial G] \asymp y|\psi'(z)|, \quad z = x + iy \in \mathbb{C}_+. \tag{4.13}$$

2.2. Maximal Functions and Luzin Function. The theory of analytic functions in G can be developed parallel to the theory of functions in $\mathbb{C}_+$ (cf. Privalov (1950)). In particular, the Hardy classes $H^p(\mathbb{C}_+)$ correspond to the Smirnov classes $E^p(G)$, $p > 0$ (cf. Sect. 8.1 of Chapter 1).

Let $f \in E^p$, $1 \le p < \infty$. We can associate with f three types of functions on the boundary ∂G of G.

(i) The *Hardy-Littlewood maximal function*

$$Mf(x) = \sup_{I \ni z} \frac{1}{|I|} \int_I |f(\zeta)||d\zeta|,$$

where the supremum runs over all arcs $I \subset \partial G$ containing z.

(ii) The *non-tangential maximal function*

$$f^*_\alpha(z) = \sup_{\zeta \in \Gamma_\alpha(z)} |f(\zeta)|.$$

(iii) The *area integral* or *Luzin function*

$$S_\alpha(f)(z) = \left(\iint_{\Gamma_\alpha(z)} |f'(\zeta)| \, d\xi d\eta \right)^{1/2}, \quad \zeta = \xi + i\eta.$$

The maximal theorem $\|Mf\|_{L^p(\partial G)} \le c\|f\|_{L^p(\partial G)}$, $1 < p < \infty$, carries over to the case of ∂G without any changes, along with its weighted analogues. In fact, the theory of the maximal function can be formulated also for so-called "spaces of homogeneous type" (Calderón (1970), Coifman and Weiss (1970)), to which category ∂G belongs too.

Despite the fact that in G we cannot use the Poisson integral, the following estimate remains in force for analytic functions.

Lemma 4.11. $f^*_\alpha(z) \le c_\alpha Mf(z)$, $z \in \partial G$.

Indeed, let $\zeta \in \Gamma_\alpha(z)$ and let the point ζ^* be symmetric to the point ζ with respect to ∂G (in the sense of Sect. 1.2 of Chapter 1). By Cauchy's formula

$$f(\zeta) = \frac{1}{2\pi i} \int_{\partial G} f(\tau) \frac{d\tau}{\tau - \zeta} = \frac{1}{2\pi i} \int_{\partial G} f(\tau) \left[\frac{\zeta - \zeta^*}{\tau - \zeta^*} \right]^m \frac{d\tau}{\tau - \zeta}$$

for any integer $m > 0$, whence

$$|f(\zeta)| \le \frac{1}{2\pi} \int_{\partial G} |f(\tau)| \frac{\rho(\zeta, \partial G)^m}{|\tau - \zeta^*|^m |\tau - \zeta|} |d\tau| \le c_\alpha Mf(z).$$

Corollary. $\|f_\alpha^*\|_{L^p(\partial G)} \le c_{\alpha,p}\|f\|_{E^p(G)}$, $p>0$.

As in the halfspace, for $p \ge 2$ this follows from the maximal theorem, and for $0<p<2$ we have to represent f as a suitable power of a function in E^2.

Now let us turn to the Luzin function $S(f)$. Consider the function $F(z) = f[\psi(z)]$ in $\mathbb{C}_+$ and its Luzin function $S(F)$. It is easy to see that

$$|F'(z)|^2\,dxdy = |f'(\zeta)|^2 d\xi d\eta, \quad \zeta = \psi(z),$$

whence, taking account of (4.12), we obtain

$$S_{\alpha_1}(F)(x) \le S_\alpha(f)(\psi(x)) \le S_{\alpha_2}(F)(x). \tag{4.14}$$

We are interested in bilateral estimates of $\|S(f)\|_{L^p(\partial G)}$ in terms of $\|f\|_{L^p(\partial G)}$, $p>0$. In view of (4.14) the estimate

$$\|S(f)\|_{L^p(\partial G)} \asymp \|f\|_{L^p(\partial G)}$$

is equivalent to the estimate

$$\int_{-\infty}^{\infty} S(F)^p(x)|\psi'(x)|\,dx \asymp \int_{-\infty}^{\infty} |F(x)|^p|\psi'(x)|\,dx \tag{4.15}$$

for $F \in H^p$, $p>0$, in the halfplane.

We know that $|\psi'| \in (\mathrm{A}_2)$. By Theorem 4.2 (Sect. 1.3) this shows that (4.15) holds for all $p \ge 2$ (cf. Sect. 4 in Chapter 1). In particular, $\|S(f)\|_{L^2(\partial G)} \asymp \|f\|_{E^2}$. But it is possible to get the L^2 estimate without having to resort to such strong tools as weighted estimates of vector valued singular integrals. In the present section we will give an "elementary" derivation of the theory, while the complete proof of (4.15) is deferred to Sect. 3.

Remark. The lower estimate

$$\int_{\partial G} S(f)^p(z)|dz| \le c_p \int_{\partial G} |f(z)|^p|dz|, \quad f \in E^p,$$

is easy to obtain for all $p>0$. For example, if $f \in E^1$ then $f = gh$, where $g \in E^2$ and $h \in E^2$ with $\|g\|_{E^2}^2 + \|h\|_{E^2}^2 \le c\|f\|_{E^1}$. Then $S(f) \le g^* S(h) + h^* S(g)$, so the E^1 estimate follows from the E^2-estimates. As in the case of the halfplane, the converse is much harder to prove.

2.3. L^2- and H^1-Estimates. As in the case of the halfplane, it is clear that

$$\int_{\partial G} S(f)^2(z)|dz| \asymp \iint_G |f'(\zeta)|^2 \rho(\zeta, \partial G)\, d\xi d\eta.$$

We have to learn to estimate this quantity by $\|f\|_{E^2}^2$. Let us pass to the halfplane $\mathbb{C}_+$ with the aid of the map ψ. In view of (4.13) it is a question of the two-sided estimate

$$\iint_{\mathbb{C}_+} |F'(z)|^2|\psi'(z)|y\,dxdy \asymp \int_{-\infty}^{\infty} |F'(x)|^2|\psi'(x)|dx$$

where $F \in H^2(\mathbb{C}_+)$. Denote the left and the right hand side by A and B respectively; it is further convenient to introduce the auxiliary function $g = F\sqrt{\psi'}$, $g \in H^2(\mathbb{C}_+)$, so that $B = \|g\|^2_{H^2}$. (ψ' is an outer function so the question of the properties $\sqrt{\psi'}$ does not arise.)

Theorem 4.2. *If* $g = F\sqrt{\psi'} \in H^2(\mathbb{C}_+)$ *then*

$$c_1 \int_{-\infty}^{\infty} |F|^2 |\psi'| dx \le \iint_{\mathbb{C}_+} |F'|^2 |\psi'| y dx dy \le c_2 \int_{-\infty}^{\infty} |F|^2 |\psi'| dx.$$

The right inequality $A \le c_2 B$ is easy to obtain. Indeed, $F = g\psi'^{-1/2}$, $F' = g'\psi'^{-1/2} - \frac{1}{2} g\psi''\psi'^{-3/2}$, so that

$$A \le \iint_{\mathbb{C}_+} |g'|^2 y dx dy + 2 \iint_{\mathbb{C}_+} |g|^2 y \left| \frac{\psi''}{\psi'} \right|^2 dx dy.$$

In view of Theorem 4.1 the first term equals $\frac{1}{4}\|g\|^2_{H^2}$, while the second term is $\iint_{\mathbb{C}_+} |g|^2 d\mu_G$ (Sect. 2.1) and does not exceed $c\|g\|^2_{H^2}$, since μ_G is a Carleson measure.

The left inequality $c_1 B \le A$ is harder to obtain. The following elementary proof has been kindly communicated to us by Paul Koosis. In view of (4.11) and Green's formula

$$B \le c_M \int_{-\infty}^{\infty} |F|^2 \mathrm{Re}\, \psi' \, dx = c_M \int_{\mathbb{C}_+} y \Delta\{|F|^2 \mathrm{Re}\, \psi'\} \, dx dy.$$

But $\mathrm{Re}\, \psi'$ is a harmonic function so that

$$\begin{aligned} |\Delta\{|F|^2 \mathrm{Re}\, \psi'\}| &= |\Delta(|F|^2) \cdot \mathrm{Re}\, \psi' + 2\nabla |F|^2 \cdot |\nabla \mathrm{Re}\, \psi'| \\ &\le |F'|^2 \cdot |\psi'| + |F'| \cdot |F| \cdot |\psi''|'. \end{aligned}$$

Hence

$$\begin{aligned} B &\le c_M \int_{\mathbb{C}_+} |F'|^2 |\psi'| \, y dx dy + c_M \int_{\mathbb{C}_+} |F'| |F| |\psi''| \, y dx dy \\ &\le c_M A + c_M \left(\int_{\mathbb{C}_+} |F'|^2 |\psi'| \, y dx dy \right)^{1/2} \left(\int_{\mathbb{C}_+} |F|^2 |\psi'| \, y \left| \frac{\psi''}{\psi'} \right|^2 dx dy \right)^{1/2} \\ &\le c_M A + c_M A^{1/2} \, \kappa(\mu_G)^{1/2} B^{1/2}. \end{aligned}$$

The last inequality follows from the fact that μ_G is Carleson, because $\|F\sqrt{\psi'}\|^2_{H^2} = \|g\|^2_{H^2} = B$. From the estimate

$$B \le c_M A + (c_M \kappa)^{1/2} \sqrt{AB}$$

we get at once the inequality $A \geq c_1 B$.

Corollary. *If $f \in E^2(G)$ then*

$$\int_{\partial G} S(f)^2 |dz| \asymp \iint_G |f'(\zeta)|^2 \rho(\zeta, \partial G)\, d\xi d\eta \asymp \int_{\partial G} |f|^2 |dz| = \|f\|^2_{E^2}.$$

As we have already remarked in Sect. 2.2, we obtain from this at once the following estimate

$$\|S(f)\|_{L^1(\partial G)} \leq c\|f\|_{E^1}, \quad f \in E^1(G).$$

In order to prove the converse estimate, we have to repeat the reasoning in Sect. 1.4 and Sect. 1.5. All changes are completely obvious. The integral representation has to be replaced by

$$F(z) = -\frac{2}{\pi i}\int_0^\infty t dt \int_{\Lambda_t} f'(\zeta) \frac{d\zeta}{(\zeta - z - 2it)^2}$$

(concerning Λ_t, cf. Sect. 2.1), that is, by

$$f(z) = \int_0^\infty dt \int_{\Lambda_t} \frac{\lambda(\zeta)}{(\zeta - z - 2it)^2} d\zeta, \tag{4.16}$$

where

$$\lambda(\zeta) = -\frac{2}{\pi i} t f'(\zeta),\ \zeta \in \Lambda_t.$$

In particular, the geometrical properties of (4.16) are the same as for (4.7), and we can state an analogue of Theorem 4.4 and 4.7.

Theorem 4.13. (i) $\|S(f)\|_{L^1(\partial G)} \asymp \|f\|_{E^1}$, $f \in E^1(G)$;

(ii) $f \in E^1(G)$ *if and only if f admits a representation in the form* (4.16) *with*

$$\int_{\partial G} |dz| \left(\int_0^\infty \frac{dt}{t^2} \int_{\Gamma(z)\cap\Lambda_t} |\lambda(\zeta)|^2 |d\zeta| \right)^{1/2} < \infty.$$

(iii) *If $f, g \in E^2(G)$, $h \in E^p(G)$ for some $p > 0$ and $h' = f'g$, then $h \in E^1(G)$ and $\|h\|_{E^1} \leq c\|f\|_{E^2}\|g\|_{E^2}$.*

So far we have successfully modeled on a Lipschitz domain G all constructions which in Sect. 1 referred to a halfplane. But the reasonings in Sect. 1.4, in particular, the duality between H^p and $H^{p'}_-$, require new tools. The further fulfillment of this program depends on the question of boundedness in $L^p(\Gamma)$ of the Cauchy integral.

2.4. Cauchy Integral. Let $f \in L^p(\partial G)$, $1 < p < \infty$. Let us consider the *Cauchy integral*

$$F(z) = \mathcal{K}f(z) = \frac{1}{2\pi i}\int_{\partial G} f(\zeta) \frac{d\zeta}{\zeta - z}, \quad z \in G.$$

This integral operator has been discussed in detail in Sect. 6 of Chapter 2. In particular, if $\mathcal{K}$ acts continuously from $L^p(\partial G)$ into $E^p(G)$, then every function $f \in L^p(\partial G)$ can be written in the form $f = f_+ + f_-$, where $f_+ \in E^p(G)$ and $f_- \in E^p(\mathbb{C}\backslash G)$ with $\|f_\pm\|_{E^p} \leq c\|f\|_{L^p}$. It follows that an analytic function f in G, $f \in E^\delta$, $\delta > 0$, belongs to E^p if and only if

$$\left|\int_{\partial G} f(z)g(z)dz\right| \leq c\|g\|_{L^{p'}}, \quad g \in E^{p'}(\mathbb{C}\backslash G). \tag{4.17}$$

The boundedness of the Cauchy integral in L^p on an arbitrary Lipschitz curve is a basic result in the theory of singular integrals. We will discuss it in detail below in Chapter 5.

In 1987 Jones and Semmes found a completely elementary proof of the boundedness of the Cauchy integral. It is sufficient to show that $\mathcal{K}$ is bounded in L^2. After that the standard techniques of Chapter 3 (with obvious modifications) give the boundedness in L^p for all p, $1 < p < \infty$.

Theorem 4.14. *The operator $\mathcal{K}$ is bounded from $L^2(\partial G)$ into $E^2(G)$ for an arbitrary Lipschitz domain G.*

The proof of Jones and Semmes has still not been published; we have learnt about it from an oral communication by Paul Koosis.

In view of Theorem 4.12 it is sufficient to prove that

$$\iint_G |F'(\zeta)|^2 \rho(z, \partial G)\, dxdy \leq c \int_{\partial G} |f(z)|^2 |dz|. \tag{4.18}$$

But

$$F'(z) = \frac{1}{2\pi i} \int_{\partial G} f(\zeta) \frac{d\zeta}{(\zeta - z)^2},$$

so (4.18) means that the integral operator T_1,

$$T_1 f(z) = \int_{\partial G} f(\zeta) \frac{\rho(z, \partial G)^{1/2}}{(\zeta - z)^2} d\zeta$$

acts from $L^2(\partial G)$ into $L^2(G)$. This is equivalent to the boundedness of the adjoint operator $T_1^* : L^2(G) \to L^2(\partial G)$ or, equivalently, to the boundedness of the operator T_2:

$$T_2 h(\zeta) = \iint_G h(z) \frac{\rho(z, \partial G)^{1/2}}{(\zeta - z)^2} dxdy, \quad \zeta \in \partial G.$$

But the kernel of T_2 is analytic in $\zeta \in \mathbb{C}\backslash G$ so that $T_2 h$ is the boundary value of a function which is analytic in $\mathbb{C}\backslash G$ and, plainly, tends to 0 as $\zeta \to \infty$. We have to estimate $\|T_2 h\|_{L^2(\partial G)} = \|T_2 h\|_{E^2(\mathbb{C}\backslash G)}$. If we apply Theorem 4.2 once more, this time to the domain $\mathbb{C}\backslash G$, we see that (4.18) is equivalent to the boundedness of the operator T_3,

$$T_3 h(\zeta) = \iint_G h(z) \frac{\rho(\zeta, \partial G)^{1/2} \rho(z, \partial G)^{1/2}}{(\zeta - z)^3}\, dxdy,$$

acting from $L^2(G)$ into $L^2(\mathbb{C}\backslash G)$. But the boundedness of T_3 is obvious! The kernel of T_3,

$$K(\zeta, z) = \frac{\rho(\zeta, \partial G)^{1/2}\rho(z, \partial G)^{1/2}}{(\zeta - z)^3}, \quad z \in G,\ \zeta \in \mathbb{C}\backslash G,$$

is uniformly summable in both arguments:

$$\sup_\zeta \iint_G |K(\zeta, z)|\, dxdy < \infty, \quad \sup_z \iint_{\mathbb{C}\backslash G} |K(\zeta, z)|\, d\xi d\eta < \infty.$$

It is well-known that this implies the L^2 boundedness (Schur's test). Let us, for example, check the first inequality. Set $\rho(\zeta, \partial G) = \delta$. Then $|\zeta - z| > \delta$ for $z \in G$ and

$$\iint_G |K(\zeta, z)|\, dxdy \le \delta^{1/2} \iint_{|z-\zeta|>\delta} \frac{|\zeta - z|^{1/2}}{|\zeta - z|^3}\, dxdy \le 4\pi.$$

Corollary. *The singular integral operator*

$$Qf(z) = \lim_{\varepsilon\to 0} \frac{1}{2\pi i} \int_{|\zeta - z|>\varepsilon} f(\zeta)\frac{d\zeta}{\zeta - z}, \quad z \in \partial G,$$

is bounded in $L^p(\partial G)$ *for* $1 < p < \infty$ *and is of weak type* $(1,1)$.

Remark 1. 1) An analysis of the proof of Theorem 4.14 reveals that $\|\mathcal{K}\| \le c(M^2+1)$. This estimate is not sharp, because there is another proof (cf. Chapter 5) which allows one to get the estimate $\|\mathcal{K}\| \le c(M^2+1)^{3/4}$, which cannot be improved.

Remark 2. It is clear that the Jones-Semmes proof works for every domain G such that in G and in $\mathbb{C}\backslash G$ one can estimate the norm of a function f in E^2 from below in terms of the norm of $S(f)$ in $L^2(\partial G)$. In the following section we will see that a sufficient condition for this is that $|\psi'| \in (\mathrm{A}_\infty)$, which again (Jerison and Kenig (1982a), (1982b)) is equivalent to the following well-known assumption about the boundary of G: for each arc $I \subset \partial G$ holds

$$|I| \le c\,\mathrm{diam}(I).$$

Is it possible to obtain in this way, a final answer, that is, is the Carleson condition on ∂G sufficient (cf. Sect. 6 of Chapter 2 and Chapter 5)? This is not known, but the just recorded fact that the estimate for $\|\mathcal{K}\|$ is not sharp shows that the proof of the estimate of $\|f\|_{E^2}$ by $\|S(f)\|_{E^2}$ requires a radical change.

2.5. L^p-Estimates. As in the case of the halfspace, Theorem 4.14 allows us to consider $S(f)$ as the norm of the vector valued integral operator

$$[Tf(z)](\zeta) = \frac{1}{2\pi i} \int_{\partial G} f(\tau)\frac{d\tau}{(\tau - \zeta)^2}, \quad \zeta \in \Gamma(z),$$

in the Hilbert space $L^2(\Gamma(z))$. The operator T is defined in the whole of $L^2(\partial G)$ and Theorem 4.12 and Theorem 4.14 show that it is bounded. Now all reasonings in Sect. 1.3 pertain to functions in G (the formal identification of $\Gamma(z)$ for different z is a minor technical complication). In exactly the same way one carries over the proof of Theorem 4.3 to G, the rôle of the representation (4.7) being played by (4.16), and in the proof we have to use the criterion (4.17). Finally, the characterization of BMO in Sect. 1.7 likewise extends to G. Here we formulate for reference the main result.

Theorem 4.15. (i) *The analytic function f in a Lipschitz domain G belongs to $E^p(G)$, $1 \le p < \infty$, if and only if $S(f) \in L^p(\partial G)$. Also*

$$\|S(f)\|_{L^p} \asymp \|f\|_{E^p}.$$

(ii) *f is in* BMO(∂G) *if and only if the measure*

$$d\mu_f = \rho(z, \partial G)|f'(z)|^2 \, dxdy$$

is a Carleson measure on G, that is,

$$\mu_f\{z \in G : \rho(z, I) \le |I|\} \le \kappa |I|$$

for every arc $I \subset \partial G$.

§3. Estimates for Distribution Functions

3.1. The Burkholder-Gundy Theorem. In Sects. 1 and 2 we derived L^p-estimates from Calderón-Zygmund theory. But Calderón-Zygmund theory does not give estimates for $0 < p \le 1$. There exist direct methods for estimating S-functions and their analogues. Various proofs and methods for estimating can be found in Chapter IV and Chapter V of Stein's book (1970a). We now consider a universal approach set forth by Burkholder and Gundy (1972) after the appearance of Stein's book.

Let f be in $H^p(\mathbb{C}_+)$, $p > 0$, and let $0 < \alpha < \beta < \infty$.

Theorem 4.16. *For $\gamma > 0$ sufficiently small*

$$\operatorname{mes}\{x \in \mathbb{R} : S_\alpha(f)(x) > 2\lambda, f^*_\beta(x) \le \gamma\lambda\} \le c\gamma^2 \operatorname{mes}\{x \in \mathbb{R} : S_\alpha(f)(x) > \lambda\}, \quad \lambda > 0.$$

Set now $E(\gamma\lambda) = \{x \in \mathbb{R} : S_\beta(f)(x) > \gamma\lambda\}$ and let χ be the characteristic function of this set.

Theorem 4.17. *For $\gamma > 0$ sufficiently small*

$$\operatorname{mes}\{x \in \mathbb{R} : f^*_\alpha(x) > 2\lambda, M\chi(x) > \tfrac{1}{2}\} \le c\gamma^2 \operatorname{mes}\{x \in \mathbb{R} : f^*_\alpha(x) > \lambda\}, \quad \lambda > 0.$$

In reality we get a much stronger estimate (cf. Sect. 3 of Chapter 3): the set $\{S_\alpha(f) > \lambda\}$ is covered by intervals $\{I_j\}$ such that

$$\operatorname{mes}\{x \in I_j : S_\alpha(f)(x) > 2\lambda, f^*_\beta(x) \le \gamma\lambda\} \le c\gamma^2|I_j|. \tag{4.19}$$

From this follows at once, exactly as in Sect. 7 of Chapter 7, a weighted estimate valid for all p.

Theorem 4.18. *If $w \in (A_\infty)$ and $p > 0$ then*

$$\|S_\alpha(f)\|_{L^p_w(\mathbf{R})} \asymp \|f^*_\beta\|_{L^p_w(\mathbf{R})} \asymp \|f\|_{H^p_w(\mathbf{C}_+)}.$$

Let us indicate the scheme of proof of Theorem 4.16 restricting ourselves to the case $f \in C^\infty$. The open set $\{S_\alpha(f) > \lambda\}$ falls into nonintersecting intervals; let $I = (a, b)$ be one of these so that certainly $S_\alpha(f)(a) \le \lambda$. Let $x \in I$ be a point such that $S_\alpha(f)(x) > 2\lambda$ but $f^*_\beta(x) \le \gamma\lambda$. Let us compare $S_\alpha(f)(x)$ with $S_\alpha(f)(a)$. By definition $S_\alpha(f)^2(x)$ is the integral of $|f'|^2$ extended over the sector $\Gamma_\alpha(x)$. The contribution of $\Gamma_\alpha(x) \cap \Gamma_\alpha(a)$ to this integral does not exeed $S_\alpha(f)^2(a) \le \lambda^2$. On the other hand, in view of (4.2), we have in $\Gamma_\alpha(x)$

$$|f'(\zeta)| \le \frac{c}{\eta} f^*_\beta(x) \le c\frac{\gamma\lambda}{\eta}, \quad \zeta = \xi + i\eta \in \Gamma_\alpha(x).$$

The set $\Gamma_\alpha(x) \backslash \Gamma_\alpha(a)$ is an inclinated halfstrip, the width of which in the horizontal direction does not exeeed $|I|$. That is, the contribution to $S_\alpha(f)^2(x)$ of the domain

$$\{\zeta : \Gamma_\alpha(x) \backslash \Gamma_\alpha(a), \eta > |I|\}$$

does not exceed

$$c^2|I| \int_{|I|}^\infty \frac{\gamma^2\lambda^2}{\eta^2} d\eta = c^2\gamma^2\lambda^2.$$

Thus, for $\gamma > 0$ sufficiently small the contribution to the integral $S_\alpha(f)^2(x)$ of the domain $\Gamma_\alpha(x) \cap \{\eta < |I|\}$ can not be less than $(2\lambda)^2 - \lambda^2 - c^2\gamma^2\lambda^2 > \lambda^2$, that is,

$$\tilde{S}(f)(x) \stackrel{\text{def}}{=} \left(\iint_{\Gamma_\alpha(x) \cap \{\eta < |I|\}} |f'(\zeta)|^2 \, d\xi d\eta) \right)^{1/2} > \lambda.$$

This determines the L^2-estimates. Let

$$E = \{x \in I : \tilde{S}(f)(x) > \lambda, \ f^*_\beta(x) \le \gamma\lambda\}.$$

We have to prove that $|E| < c\gamma^2|I|$. But

$$\lambda^2|E| \le \int_E \tilde{S}^2 \, dx \le \iint_W \eta |f'(\zeta)|^2 \, d\xi d\eta$$

where

$$W = \bigcup_{x \in E} \{\zeta \in \Gamma_\alpha(x),\ \eta < |I|\}.$$

It is clear that W is a Lipschitz domain with the boundary not longer than $c_\alpha |I|$. By Green's formula

$$\iint_W \eta |f'|^2 \, d\xi d\eta = \frac{1}{4}\int_{\partial W} \eta \frac{\partial |f|^2}{\partial n} |d\zeta| - \frac{1}{4}\int_{\partial W} |f|^2 \frac{\partial \eta}{\partial n} |d\zeta|,$$

where $\frac{\partial}{\partial n}$ is differentiation in the direction of the outer normal of ∂W. But $\eta < |I|$, $|\frac{\partial \eta}{\partial n}| \le 1$ and throughout W we have $|f(\zeta)| \le f^*_\beta(x) \le \gamma\lambda$ and $|f'(\zeta)| \le c\frac{\gamma\lambda}{\eta}$. Hence

$$\iint_W \eta |f'|^2 \, dxdy \le c\gamma^2\lambda^2|I|,$$

from which the desired estimate $|E| \le c\gamma^2|I|$ follows.

The proof of the converse of Theorem 4.17 is similar but more tedious, because using $S(f)$ one can estimate only the derivative of f, so that one has to prove in addition that $|f(\zeta)| \le \lambda + c\gamma\lambda$ for $\zeta \in \Gamma_\alpha(x)$ and $\eta = |I|$.

3.2. Dahlberg's Theorem. Dahlberg (1980) proved an analogue of Theorem 4.18 for functions in a Lipschitz domain.

Theorem 3.2. *Let G be a planar Lipschitz domain, w a weight on ∂G satisfying the (A_∞) condition, $p > 0$, $\alpha > 0$, $\beta > 0$ and z_0 a fixed point of G. If f is an analytic function in G with $f(z_0) = 0$ then*

$$\|S_\alpha(f)\|_{L^p_w(\partial G)} \asymp \|f^*_\beta\|_{L^p_w(\partial G)} \asymp \|f\|_{E^p_w(\partial G)}.$$

Remark. 1) In a special Lipschitz domain (Sect. 2.1) one can take $z_0 = \infty$.

2) In fact, Dahlberg (1980) proved his theorem directly for harmonic functions in $\mathbb{R}^n$. We will review the corresponding construction in Sect. 4.

Dahlberg's proof follows the scheme of the proof of Theorem 4.18 of Burkholder and Gundy. But we can not use Green's formula in G, as we do not possess an analogue of the special function η.

We will not describe the proof for a Lipschitz domain. Instead we shall show how using Dahlberg's method one can conclude the proof of Theorem 4.16. In the notation of Sect. 3.1 we have to estimate the measure of the set E. Now W is Lipschitz with a Lipschitz constant depending only on α. Let us introduce the point $\zeta_0 = \xi_0 + i\eta_0 \in W$, $\eta_0 > |I|/2$, and let us consider the harmonic measure ω of W representing ζ_0 and Green's function $\mathcal{G}$ of W with pole at ζ_0. Both $\mathcal{G}$ and ω have well known expressions in terms the conformal mapping of W onto the disk $\mathbb{D}$ or the halfplane $\mathbb{C}_+$. In Sect. 2.1 we discussed the properties of these mappings, from which follows the following:

(i) $\omega[B(z,2r)] \le c\omega[B(z,r)]$, $z \in \partial W$, $r > 0$.
(ii) $\mathcal{G}(\zeta) \asymp \omega[B(\zeta, 2\rho(\zeta, \partial W))]$, $\zeta \in W$.
(iii) The arc length of ∂W satifies condition (A_2) with respect to ω.
Now we can write

$$\int_E \tilde{S}^2 d\omega(x) \le \iint_W |f'(\zeta)|^2 \omega[B(\zeta, 2\rho(\zeta, \partial W))]\, d\xi d\eta$$
$$\le c \iint_W |f'(\zeta)|^2 \mathcal{G}(\zeta)\, d\xi d\eta = \frac{c}{4} \iint_W \Delta(|f|^2)\mathcal{G}(\zeta)\, d\xi d\eta.$$

But in view of Green's formula

$$\iint_W \Delta(|f|^2)\mathcal{G}(\zeta)\, d\xi d\eta = \int_{\partial W} |f|^2 d\omega + |f(\zeta_0)|^2 \le c^2\gamma^2\lambda^2$$

so that

$$\lambda^2\omega(E) \le c^2\gamma^2\lambda^2, \quad \omega(E) \le c^2\gamma^2.$$

From condition (A_2) on the arc length we conclude that

$$|E| \le C\gamma^\delta |I|, \quad \delta > 0.$$

This is the estimate which we need for $|E|$, with a slightly deteriorated dependence on γ.

Thus, in Dahlberg's proof, the function η is replaced by Green's function $\mathcal{G}$, but in addition one has to integrate $\tilde{S}$ not with respect to arc length but with respect to the harmonic measure ω and one gets an estimate for $\omega(E)$, not $|E|$. In order to return to $|E|$ one has to invoke the Muckenhoupt condition, that is, a rather complicated technique.

3.3. The g-Function. The Luzin function $S(f)$ is a deep and most useful object in Littlewood-Paley theory, but not the only one. Earlier Littlewood and Paley had considered another quadratic expression

$$g(f)(x) = \left(\int_0^\infty |f'(x+iy)|^2 y dy\right)^{1/2}, \quad x \in \mathbb{R},$$

where f is as before an analytic function in the upper halfplane $\mathbb{C}_+$.

Besides $g(f)$ let us consider the more general *Littlewood-Paley function*

$$g_\lambda(f)(x) = \left(\iint_{\mathbb{C}_+} |f'(\zeta)|^2 \left(\frac{\eta}{|\xi - x| + \eta}\right)^{2\lambda} d\xi d\eta\right)^{1/2}, \quad x \in \mathbb{R},$$

where $\lambda > 1$.

In the same way as $S(f)$ controls the nontangential convergence of $f(\zeta)$ to $f(x)$, $g(f)(x)$ controls vertical convergence $f(x+iy) \to f(x)$, and $g_\lambda(f)(x)$

convergence under arbitrary approach $\zeta \to x$. From the mean value theorem we readily find the estimates

$$g(f) \leq c_\alpha S_\alpha(f)$$

and

$$S_\alpha(f) \leq c_{\lambda,\alpha} g_\lambda(f).$$

Moreover, it is easy to see that $g_\lambda \leq g_\mu$ for $\lambda \geq \mu$.

A computation of the L^2 norms of $g(f)$ and $g_\lambda(f)$ immediately leads to the same expression

$$\iint_{\mathbf{C}_+} |f'|^2 y dx dy = \frac{1}{4} \int_{-\infty}^{\infty} |f|^2 dx,$$

so that

$$\|g(f)\|_{L^2} \asymp \|g_\lambda(f)\|_{L^2} \asymp \|f\|_{H^2}.$$

As in Sect. 1.3 it is easy to get a two-sided estimate

$$\|g(f)\|_{L^p_w} \asymp \|f\|_{H^p_w}$$

in the hypothesis that $1 < p < \infty$ and $w \in (\mathrm{A}_p)$. However, for $g_\lambda(f)$ one can give estimates for certain $p \leq 1$.

Theorem 4.20. *For* $2/\lambda < p < \infty$

$$\|g_\lambda(f)\|_{L^p(\mathbf{R})} \asymp \|f\|_{H^p(\mathbf{C}_+)}.$$

In one direction ($\|f\| \leq \|g_\lambda(f)\|$) this estimate is obvious, because even $\|S(f)\|_{L^p} \geq \|f\|_{H^p}$ for all $p > 0$.

The estimate of $g_\lambda(f)$ from above can be obtained in various ways (cf. Aguilera and Segovia (1977), Stein (1970a)). Here we describe a result of Muckenhoupt and Wheeden (1974), who proved Theorem 4.20 following the route taken by Burkholder and Gundy (Sect. 3.1).

In order to estimate $G_\lambda(f)$ we require a *maximal function* of a new kind (cf. Sect. 1.7). Set

$$T_{\lambda r}(f)(x) = \sup_{I \ni x} \left(\frac{1}{|I|^\lambda} \iint_{\square(I)} |f(\zeta)|^r \eta^\lambda \, d\xi d\eta \right)^{1/2},$$

where $\lambda \geq 1$ while $r > 0$. Note that always

$$f^*_\alpha(x) \leq c_{\alpha\lambda r} T_{\lambda r}(f)(x).$$

Lemma 4.21. *If* $1 < \lambda < r < \infty$ *then*

$$T_{\lambda r}(f)(x) \leq c_{\lambda r} M_{r/\lambda}(f)(x), \quad x \in \mathbb{R}.$$

The proof of this lemma can be found in Muckenhoupt and Wheeden (1974); it is based on the Hardy-Littlewood-Sobolev inequality for fractional integrals.

Corollary 1. *If* $1 < \lambda < r$, $r/\lambda < p < \infty$ *and* $w \in (\mathrm{A}_{p\lambda/r})$ *then*

$$\int_{-\infty}^{\infty} T_{\lambda r}(f)^p w dx \le c \int_{-\infty}^{\infty} |f|^p w dx.$$

Corollary 2. *If* $w \in (A_1)$ *then*

$$w\{x : T_{\lambda r}(f)(x)\} \le cA^{-r/\lambda} \int_{-\infty}^{\infty} |f|^{r/\lambda} w dx, \quad A > 0.$$

Let now $f \in H^p(\mathbb{C}_+)$, $p > 0$. For every $s > 1$ there exists a function $h \in H^s(\mathbb{C}_+)$ such that $|f|^p = |h|^s$ on $\mathbb{R}$. By Lemma 4.21

$$T_{\lambda r}(f) \le [T_{\lambda, rs/p}(h)]^{s/p} \le c[M_{rs/\lambda p}(h)]^{s/p}.$$

If s is suffciently large we can apply Corollary 1. We state the result at once in the weighted case.

Lemma 4.22. *Let* $\lambda > 1$, $0 < r < \infty$.
(i) For $r/\lambda < p < \infty$ *and* $w \in (A_{p\lambda/r})$

$$\|T_{\lambda r}(f)\|_{L^p_w(\mathbf{R})} \le c\|f\|_{H^p_w}, \quad f \in H^p_w(\mathbb{C}_+).$$

(ii) For $w \in (A_1)$

$$w\{x : T_{\lambda r}(f)(x) > A\} \le cA^{-r/\lambda}\|f\|^{r/\lambda}_{H^{r/\lambda}_w}, \quad A > 0.$$

It is obvious that the Lemma 4.2 extends Corollary 1 and 2 to the case $\lambda \ge r$. In the following we need only the case $r = 2$ but for the proof it is necessary to consider arbitrary r.

Theorem 4.23 (Muckenhoupt-Wheeden). *Let* $\lambda > 1$ *and the function* f *be analytic in* $\mathbb{C}_+$. *Then for* $\gamma > 0$ *sufficiently small*

$$\mathrm{mes}\{x \in \mathbb{R} : g_\lambda(f)(x) > 2A,\, T_{\lambda 2}(f)(x) \le \gamma\} \le c\gamma^2 \mathrm{mes}\{x \in \mathbb{R} : g_\lambda(f)(x) \ge A\}, \quad A > 0.$$

Now Theorem 4.20 is an immediate consequence. Moreover, we get, as usual, a weighted analogue.

Theorem 4.24. *For* $\lambda > 1$, $2/\lambda < p < \infty$ *and* $w \in (A_{p\lambda/2})$

$$\|g_\lambda(f)\|_{L^p_w(\mathbf{R})} \le c\|f\|_{H^p_w(\mathbb{C}_+)}.$$

§4. Harmonic Functions

4.1. The Functions of Littlewood-Paley and Luzin. In Sects. 1–3 Littlewood-Paley theory was developed for analytic functions in the halfspace (disk) or in planar Lipschitz domains. But the whole theory extends to harmonic functions in $\mathbb{R}^n$. In fact, many of the constructions, for instance the approach of Dahlberg, arose precisely in a desire to rid oneself from "planar" pecularities.

Let f be a function in $\mathbb{R}^n$ and let us assume, to fix the ideas, that $f \in L_p(\mathbb{R}^n)$, $1 \le p \le \infty$. With f we associate its Poisson integral in $\mathbb{R}^{n+1}_+$ (Sect. 7.1 of Chapter 1)

$$u(x,t) = P_t * f(x) = \mathcal{P}_t f(x), \quad x \in \mathbb{R}^n,\ t > 0;$$

let us consider its gradient

$$\nabla u(x,t) = \left(\frac{\partial u}{\partial t}, \frac{\partial u}{\partial x_1}, \dots, \frac{\partial u}{\partial x_n}\right).$$

The *Littlewood-Paley function* $g(f)$ or $g(u)$ is defined by the formula

$$g(f)(x) = g(u)(x) = \left(\int_0^\infty t|\nabla u(x,t)|^2\,dt\right)^{1/2}, \quad x \in \mathbb{R}^n,$$

and the *truncated Littlewood-Paley function* $g_0(f)$ or $g_0(u)$ by

$$g_0(f)(x) = g_0(u)(x) = \left(\int_0^\infty t\left|\frac{\partial u}{\partial t}\right|^2 dt\right)^{1/2}, \quad x \in \mathbb{R}^n.$$

Clearly $g_0 \le g$.

Let $\lambda > 1$. The *Littlewood-Paley function* $g_\lambda(f)$ or $g_\lambda(u)$ is defined by the formula

$$g_\lambda(f)(x) = g_\lambda(u)(x) = \left(\iint_{\mathbb{R}^{n+1}_+} t^{1-n}|\nabla u(y,t)|^2\left(\frac{t}{|x-y|+t}\right)^{\lambda n} dydt\right)^{1/2}, x \in \mathbb{R}^n.$$

Finally, we define the *Luzin function* $S_\alpha(f)$ or $S_\alpha(u)$, $\alpha > 0$, as

$$S_\alpha(f)(x) = S_\alpha(u)(x) = \left(\iint_{\Gamma_\alpha(x)} t^{1-n}|\nabla u(y,t)|^2\,dydt\right)^{1/2}, \quad x \in \mathbb{R}^n,$$

where $\Gamma_\alpha(x) = \{(y,t) : |y-x| < \alpha t\}$ is the *Luzin cone* .

If $n = 1$ and $f \in H^p$ then all these definitions reduce to the ones in Sect. 1 and Sect. 3. The Luzin function will as before be referred to as the area function, although now it is not connected with any area whatsoever.

The simplest inequalities between $g(u)$, $g_\lambda(u)$, $S_\alpha(u)$ and the non-tangential maximal function u^*_β, mentioned in Sects. 3.1 and 3.3 are based only on the mean value theorem and so remain in force also in the multivariate theory.

In Coifman, Meyer and Stein (1985) the analogue of the S-function for an arbitrary function u in $\mathbb{R}^{n+1}_+$ is considered. For example, it is shown there that for any differentiable function $u(x,t)$ we have

$$\|S_\alpha(u)\|_{L_p(\mathbf{R}^n)} \le c_{\alpha\beta p}\|S_\beta(u)\|_{L_p(\mathbf{R}^n)}$$

for all $\alpha > 0$, $\beta > 0$ and $0 < p < \infty$. For $p = \infty$ this is not true.

4.2. The Main Estimates in $L_p(\mathbb{R}^n)$. The L^p-estimates for $g(f)$ and $S(f)$ are the same as in the case of analytic functions.

Theorem 4.25. *For $1 < p < \infty$ and $w \in (A_p)$*

$$\|g(f)\|_{L^p_w(\mathbf{R}^n)} \asymp \|g_0(f)\|_{L^p_w(\mathbf{R}^n)} \asymp \|S_\alpha(f)\|_{L^p_w(\mathbf{R}^n)} \asymp \|f\|_{L^p_w(\mathbf{R}^n)}.$$

Theorem 4.26. *For any $\alpha, \beta > 0$, $0 < p < \infty$ and $w \in (A_\infty)$*

$$\|S_\alpha(f)\|_{L^p_w(\mathbf{R}^n)} \asymp \|u^*_\beta\|_{L^p_w(\mathbf{R}^n)} \asymp \|f\|_{H^p_w(\mathbf{R}^n)}.$$

Theorem 4.27. *For $2/\lambda < p < \infty$ and $w \in (A_{p\lambda/2})$*

$$\|g_\lambda(f)\|_{L^p_w(\mathbf{R}^n)} \le c\|f\|_{H^p_w(\mathbf{R}^n)}.$$

The plan of the proof of Theorems 4.25–4.27 is the same as for the corresponding results of Sect. 1 and Sect. 3. Let us recall the main stages.

(i) The integral of the square of any of the g- or S-functions over the whole space $\mathbb{R}^n$ is proportional to

$$A(u) = \iint_{\mathbf{R}^{n+1}_+} t|\nabla u(x,t)|^2\, dxdt = \frac{1}{2}\|f\|^2_{L^2(\mathbf{R}^n)}.$$

This follows, as in Sect. 1.2, either from Green's formula, applied in $\mathbb{R}^{n+1}_+$ to the functions $|u|^2$ and t, or from Plancherel's formula applied to the convolution $u = P_t * f$. Concerning the truncated function $g_0(f)$, Plancherel's formula gives for it

$$\|g_0(f)\|^2_{L^2} = \iint_{\mathbf{R}^{n+1}_+} t\left|\frac{\partial u}{\partial t}\right|^2 dxdt = \frac{1}{4}\|f\|^2_{L^2}.$$

(ii) As soon as the L^2-estimates have been obtained, Calderón-Zygmund theory gives us Theorem 4.25 exactly as in Sect. 1.3. The only difference is that we have to replace the kernel $K(x)$ in Sect. 1.3 by the vector valued kernel

$$[K_t(x)](y,t) = \nabla P_t(x-y), \quad (y,t) \in \Gamma(x),$$

where P_t is the Poisson kernel in $\mathbb{R}^{n+1}_+$. The estimates of the kernel are thereby not changed.

(iii) The Burholder-Gundy estimates for the distribution functions

$$\text{mes}\{x : S_\alpha(f)(x) > 2\lambda, u^*_\beta \le \gamma\lambda\} \le c\gamma^2 \text{mes}\{x : S_\alpha(f)(x) > \lambda\},$$

for $0 < \alpha < \beta$, remain in force completely, along with their proof in Sect. 3.1, which was based only on properties of harmonic functions. The open set $\{S_\alpha(f) > \lambda\}$ is decomposed into Whitney balls (Sect. 3.3 of Chapter 1) $\{B_j\}$. If B is any of these balls with radius r, there exists a point a, not further away from B than $4r$, such that $S_\alpha(f)(a) \le \lambda$. The comparison of $S_\alpha(f)(x)$ and $S_\alpha(f)(a)$ goes as in Sect. 3.1. Of course, the set $\Gamma_\alpha(x)\backslash\Gamma_\alpha(a)$ is not anymore a strip, but the measure of its horizontal section at height t does not exceed $c \cdot |x-a| \cdot t^{n-1}$, which is sufficient for the estimates. The proof of is completed by applying Green's formula.

(iv) The estimates of the functions $g_\lambda(f)$ are obtained as in Sect. 3.3 by comparing them with the maximal functions

$$T_{\lambda r}(u)(x) = \sup_{B \ni x} \left(\frac{1}{|B|^\lambda} \int_{\square(B)} |u(y,t)|^r \, t^{(\lambda-1)n+1} \, dydt \right)^{1/r}.$$

Muckenhoupt and Wheeden's proof (1974) of the estimates for the distribution functions (Theorem 4.23) carries immediately over to the n-dimensional case, because everything blows down to estimates for $T_{\lambda r}$. The reasonings in Sect. 3.3 are still available. The work with H^p is scarcely more complicated: by Sect. 8.2 of Chapter 1 one can find a positive harmonic function h such that $|u|^p \le h^s$ and $\|h^s\|_{L^1(\mathbb{R}^n)} \le c\|f\|^p_{H^p(\mathbb{R}^n)}$. No other changes are needed.

Next, let us turn to the estimates in $\text{BMO}(\mathbb{R}^n)$.

Theorem 4.28. $f \in \text{BMO}(\mathbb{R}^n)$ *if and only if the measure* μ_f,

$$d\mu_f(x,t) = t|\nabla u(x,t)|^2 \, dxdt,$$

is a Carleson measure in $\mathbb{R}^n_+$, *and then*

$$\kappa(\mu_f) \overset{\text{def}}{=} \sup \frac{\mu_f(\square(B))}{|B|} \asymp \|f\|^2_{\text{BMO}}.$$

The proof can be found in Coifman, Meyer, and Stein (1985), Fefferman and Stein (1972); it is analogous to the reasonings in Sect. 1.7.

As in the one-dimensional case, the measure $|\nabla u(x,t)|\,dxdt$ is not by necessity Carleson. But, according to Garnett (1981), it is always possible, given $f \in \mathrm{BMO}(\mathbb{R}^n)$, to find a function $U \in C_0^\infty(\mathbb{R}_+^{n+1})$ such that
(i) $\sup_{t>0} |U(x,t)| \in L^1_{\mathrm{loc}}(\mathbb{R}^n)$;
(ii) $\lim_{t\to 0} U(x,t) = f(x)$ a. e.;
(iii) the measure $|\nabla U(x,t)|\,dxdt$ is Carleson in $\mathbb{R}_+^{n+1}$.

4.3. Lipschitz Domains. In Sect. 2, while studying functions in planar Lipschitz domains, we used conformal mapping onto the halfplane, a powerful tool, which completely fails in the multidimensional case. Nevertheless, after the work of Carleson (1967), Hunt and Wheeden (1968) and, in particular, Dahlberg (1977), (1980), it became clear that the Littlewood-Paley theory could be carried over to harmonic functions in Lipschitz domains in $\mathbb{R}^n$.

Let now G be a bounded Lipschitz domain in $\mathbb{R}^n$ that is starshaped with respect to the origin $0 \in \mathbb{R}^n$. The *Luzin cone* $\Gamma_\alpha(x)$, $x \in \partial G$, $\alpha > 0$, is defined by the formula

$$\Gamma_\alpha(x) = \{y \in G : |y - x| < (1+\alpha)\rho(y, \partial G)\}.$$

For harmonic functions in G we can again define the *non-tangential maximal function*

$$u_\alpha^*(x) = \sup_{y\in\Gamma_\alpha(x)} |u(y)|$$

and the *Luzin function*

$$S_\alpha(u)(x) = \left(\int_{\Gamma_\alpha(x)} |\nabla u(y)|^2 \rho(y, \partial G)^{2-n}\,dy\right)^{1/2}.$$

Let σ be the Lebesgue measure on ∂G, that is the $(n-1)$-dimensional Hausdorff measure; with respect to it we can define the Hardy-Littlewood maximal functions $M_p f$, $p > 0$, and the Muckenhoupt conditions $(\mathrm{A}_p)(d\sigma)$ and $(\mathrm{A}_\infty)(d\sigma)$.

As in the case of the halfplane, it is easy to verify that

$$\|u_\alpha^*\|_{L^p(d\sigma)} \asymp \|u_\beta^*\|_{L^p(d\sigma)}, \quad p > 0.$$

Let now ω be harmonic measure on ∂G, representing the point 0, and let $\mathcal{G}$ be Green's function with pole at 0. Let us recall their principal properties, the proof of which can be found in Hunt and Wheeden (1968) and Dahlberg (1977) (cf. further Jerison and Kenig (1982b)).

Lemma 4.29. $\omega[B(x,2r)] \le C\omega[B(x,r)]$, $x \in \partial G$, $r > 0$.

Lemma 4.30. *Uniformly in* $x \in G\backslash\frac{1}{2}G$ *holds*

$$\omega[B(x, 2\rho(x,\partial G))] \asymp \rho(x,\partial G)^{n-2} \cdot \mathcal{G}(x).$$

Lemma 4.31. *The measures σ and ω are mutually absolutely continuous and the derivative $k = d\omega/d\sigma$ satisfies*

$$\frac{1}{\sigma(B)}\int_B k^2 d\sigma \le c\left(\frac{1}{\sigma(B)}\int_B k d\sigma\right)^2$$

for each ball B with center on ∂G.

Corollary. *The measure σ satisfies Muckenhoupt's (A_2) condition with respect to ω: $\sigma \in (\mathrm{A}_2)(d\omega)$.*

Lemma 4.32. *If u is harmonic in G and continuous on $\bar{G}$ then*

$$u_\alpha^*(x) \le c_\alpha \tilde{M} u(x), \quad \alpha > 0, x \in \partial G,$$

where $\tilde{M}$ is the Hardy-Littlewood maximal function with respect to the harmonic measure ω.

Corollary. *For $2 \le p < \infty$ we have $\|u_\alpha^*\|_{L^p(d\sigma)} \le c_p \|u\|_{L^p(d\sigma)}$.*

The constants in the Lemmas 4.29–4.32 depend on the dimension of the space and the Lipschitz constant of G.

Dahlberg (1980) has proved the following theorem for Lipschitz domains in $\mathbb{R}^n$.

Theorem 4.33. *Let $\alpha > 0$, $\beta > 0$, $p > 0$ and let w be a weight function on ∂G satisfying the (A_∞) condition with respect to σ. If u is harmonic in G with $u(0) = 0$ then*

$$\|u_\alpha^*\|_{L_w^p(d\sigma)} \asymp \|S_\beta(u)\|_{L_w^p(d\sigma)}.$$

The plan of Dahlberg's proof has already been discussed in Sect. 3.2. Of course, in $\mathbb{R}^n$ one has to utilize the estimates for $\mathcal{G}$ and ω in Lemma 4.29–4.32.

Corollary 1. *For $2 \le p < \infty$*

$$\|S_\alpha(u)\|_{L^p(d\sigma)} \asymp \|u\|_{L^p(d\sigma)}.$$

Corollary 2.

$$\int_G |\nabla u(y)|^2 \rho(y, \partial G) dy \asymp \int_{\partial G} |u(x)|^2 d\sigma(x).$$

§5. Non-Poisson Means

5.1. Main Constructions. It is of great interest to extend Littlewood-Paley theory to the case when instead of the Poisson integral one considers other ways of averaging functions. Here we consider convolutions of functions in $\mathbb{R}^n$ with a smooth kernel other than the Poisson kernel.

Let $\varphi \in L^1(\mathbb{R}^n) \cap C^2(\mathbb{R}^n)$ be such that

$$|\varphi(x)| \le \frac{1}{1+|x|^{n+1}}, \quad |\nabla\varphi(x)| \le \frac{1}{1+|x|^{n+2}}, \quad |\nabla^2\varphi(x)| \le \frac{1}{1+|x|^{n+3}}. \tag{4.20}$$

Let us consider the kernel

$$\varphi_t(x) = t^{-n}\varphi(x/t), \quad t>0,\, x \in \mathbb{R}^n,$$

and the convolution operator with this kernel

$$u(x,t) = P_t f(x) \stackrel{\text{def}}{=} \varphi_t * f(x).$$

Here $f \in L^p(\mathbb{R}^n)$, $1 \le p \le \infty$. Let us list some simple properties of the operators P_t.

(i) Set

$$M^{(t)} f(x) = \sup \frac{1}{|B|}\int_B |f|,$$

with the supremum extended over all balls $B \ni x$ of radius $> t$. Then

$$|u(x,t)| \le cM^{(t)} f(x) \le cMf(x).$$

(ii) In particular,

$$u_\alpha^*(x) = \sup_{\Gamma_\alpha(x)} |u(y,t)| \le cMf(x).$$

(iii) If μ is a Carleson measure in $\mathbb{R}_+^{n+1}$ then

$$\int_{\mathbb{R}_+^{n+1}} |u(x,t)|^p d\mu(x,t) \le c_p \int_{\mathbb{R}^n} |f|^p, \quad 1<p<\infty.$$

Next let us construct the analogue of the *Luzin function*. The definition of the S function does not involve the Poisson kernel itself but its derivative. Therefore, let ψ be another function satisfying (4.20), also satisfying

$$\int_{\mathbb{R}^n} \psi(x)dx = 0. \tag{4.21}$$

Consider the corresponding convolution operators

$$v(x,t) = Q_t f(x) = \psi_t * f(x), \quad t>0,\, x\in\mathbb{R}^n.$$

In particular, if we take

$$\psi(x) = n\varphi(x) + x \cdot \nabla\psi(x), \tag{4.22}$$

then

$$Q_t f(x) = -t\frac{\partial}{\partial t}P_t f(x), \tag{4.23}$$

but such a choice is not necessary.

We can now define the analogue of the *Luzin function*

$$S_\alpha(f)(x) = \left(\int_{\Gamma_\alpha(x)} |Q_t f(y)|^2 t^{-n-1}\,dydt\right)^{1/2}, \quad x \in \mathbb{R}^n.$$

Remark. If u is the Poisson integral, then in view of (4.23) v does not correspond to ∇u but to $t\nabla u$. This explains the difference in the definition of $S(f)$ here and in Sect. 4.

As in Sect. 1, we require, besides convolutions expressing v in terms of f, also potentials allowing us to write down the converse operation.

Let $w(x,t)$ be a function in $\mathbb{R}^{n+1}_+$, which also will be written $w_t(x)$. Consider the potential

$$h(x) = \iint_{\mathbb{R}^{n+1}_+} \psi_t(x-y)w(t,y)\frac{dydt}{t} = \int_0^\infty Q_t w_t \frac{dt}{t}. \tag{4.24}$$

For the function w we can consider an analogue of the *Luzin function*

$$A_\alpha(w)(x) = \left(\int_{\Gamma_\alpha(x)} |w(y,t)|^2 \frac{dydt}{t^{n+1}}\right)^{1/2}.$$

Then, if for example $f \in \mathcal{D}(\mathbb{R}^n)$,

$$\int_{\mathbb{R}^n} f(x)h(x)dx = \iint_{\mathbb{R}^{n+1}_+} v(y,t)w(t,y)\frac{dydt}{t} \tag{4.25}$$

and, in particular, in view of Schwarz's inequality,

$$\left|\int_{\mathbb{R}^n} f(x)h(x)dx\right| \le c_a \int_{\mathbb{R}^n} S_\alpha(f)(x)A_\alpha(w)(x)dx. \tag{4.26}$$

The formulae (4.25) and (4.26) allow us to apply duality considerations while estimating $S(f)$ by f, on the one hand, or h by $A(w)$, on the other.

Finally, let us give the analogue of the integral representation (4.6). To this end, let us assume that ψ is real and radially symmetric, that is, $\psi(x)$ depends only on $|x|$. Then $\hat{\psi}_t(\xi) = \hat{\psi}(t\xi)$ too depends on $|\xi|$ only and the Fourier transform of the convolution $Q_t^2 f(x)$ in the variable x equals $\hat{\psi}_t(\xi)^2 \hat{f}(\xi)$. Let now

$$h(x) = \int_0^\infty Q_t^2 f(x) \frac{dt}{t}.$$

Then

$$\hat{h}(\xi) = \int_0^\infty |\hat{\psi}(t\xi)|^2 \hat{f}(\xi) \frac{dt}{t} = \hat{f}(\xi) \int_0^\infty |\hat{\psi}(t\xi)|^2 \frac{dt}{t}.$$

The obvious change of variables $t \mapsto t/|\xi|$ reveals that the last integral does not depend on ξ and equals a positive constant $c_0 = c_0(\psi)$. (The integral is convergent because $\hat{\psi}(0) = 0$ in view of (4.21), while in view of (4.20) $\hat{\psi}$ must be a smooth function.) Thus $h = c_0 f$ and

$$f(x) = c_0^{-1} \int_0^\infty Q_t^2 f(x) \frac{dt}{t}, \tag{4.27}$$

that is, f admits a representation as a potential of the type (4.24), with

$$w_t(x) = c_0^{-1} Q_t f(x) = c_0^{-1} v(x,t).$$

In particular, $A(w) = c_0^{-1} S(f)$. Thus we have the full analogue of the representation (4.6), to which (4.27) reduces if we let φ be the Poisson kernel, defining ψ according to (4.22).

Remarks. 1) If the kernel ψ is not radially symmetric, the question of representing an arbitrary functions as a potential (4.24) becomes harder. In the book Folland and Stein (1982) the question of such a representation is discussed for other special kernels.

2) Of course, the integral (4.27) can be interpreted literally only for smooth functions f, e. g. for $f \in \mathcal{D}(\mathbf{R}^n)$. In the general case we have to consider sections

$$f_\varepsilon(x) = c_0^{-1} \int_\varepsilon^{1/\varepsilon} Q_t^2 f \frac{dt}{t}$$

and study specially the character of their convergence to $f(x)$ as $\varepsilon \to 0$. Such a study may be quite involved, but here we will not enter into it, always assuming that we take the estimates for $f \in \mathcal{D}$. For details we refer to David (1986), Folland and Stein (1982).

5.2. L^p-Estimates. It is clear that

$$\int_{\mathbf{R}^n} S_\alpha(f)^2(x) = c_\alpha \iint_{\mathbf{R}_+^{n+1}} |Q_t f(x)|^2 \frac{dx dt}{t}.$$

Now, in view of Plancherel's formula

$$\int_{\mathbf{R}^n} |Q_t f|^2 dx = \int_{\mathbf{R}^n} |\hat{\psi}(t\xi)|^2 |\hat{f}(\xi)|^2 d\xi,$$

whence

$$\iint_{\mathbf{R}_+^{n+1}} |Q_t f|^2 \frac{dxdt}{t} = \int_{\mathbf{R}^n} |\hat{f}(\xi)|^2 d\xi \int_0^\infty |\hat{\psi}(t\xi)|^2 \frac{dt}{t} = c_0 \int_{\mathbf{R}^n} |\hat{f}(\xi)|^2 d\xi.$$

This means that

$$\int_{\mathbf{R}^n} S_\alpha(f)^2(x)\, dx = c_\alpha c_0 \int_{\mathbf{R}^n} |f(x)|^2 dx,$$

i. e.

$$\|S(f)\|_{L^2} = c\|f\|_{L^2}.$$

As in Sect. 1, it is easy to obtain from this L^p-estimates via Calderón-Zygmund theory. The assumptions imposed on ψ guarantee that the estimates needed hold for the kernel of the integral operator in view. The L^2-isometry further gives us the inverse estimates and we arrive thus at the following theorem.

Theorem 4.34. *If* $1 < p < \infty$ *then*

$$\|S_\alpha(f)\|_{L^p(\mathbf{R}^n)} \asymp \|f\|_{L^p(\mathbf{R}^n)}$$

for all $\alpha > 0$.

The duality formulae (4.25) and (4.27) give at once a theorem on integral representation.

Theorem 4.35. *If the function* ψ *is radially symmetric, then* $f \in L^p(\mathbb{R}^n)$, $1 < p < \infty$, *if and only if* f *admits a representation as a potential* (4.24) *with* $\|A_\alpha(w)\|_{L^p(\mathbf{R}^n)} < \infty$.

Further information on L^p-estimates can be found in the book Folland and Stein (1982). In particular, there the extension to H^p, $0 < p \le 1$, is discussed (this is possible in supplementary assumptions on ψ) and furthermore the dependence on the parameter α and the analogue of the function $g_\lambda(f)$ are treated.

The integral representation (4.27) gives a new way of estimating linear operators in L^2. Let T be a linear operator acting, for instance, from $\mathcal{D}(\mathbb{R}^n)$ into $\mathcal{D}'(\mathbb{R}^n)$. We wish to prove that it is continuous in L^2, that is the estimate

$$|\langle Tf, g\rangle| \le c\|f\|_{L^2}\|g\|_{L^2}, \quad f, g \in \mathcal{D}(\mathbb{R}^n).$$

Let $\psi \in \mathcal{D}(\mathbb{R}^n)$ be radially symmetric satisfying (4.21). Let us replace the functions f and g by the integral representations (4.27). Assuming that all passages to the limit are permissible, we obtain

$$\begin{aligned}\langle Tf, g\rangle &= c_0^{-2} \iint_0^\infty \langle Q_s T Q_t \cdot Q_t f, Q_s g\rangle \frac{dtds}{ts} \\ &= c_0^{-2} \iint_{\mathbf{R}_+^{n+1} \times \mathbf{R}_+^{n+1}} L(x,t;y,s) Q_t f(x) Q_s g(y) \frac{dxdt}{t}\frac{dyds}{s},\end{aligned}$$

where

$$L(x,t;y,s) = \langle T\psi_t^x, \psi_s^y \rangle, \quad \psi_t^x(z) \stackrel{\text{def}}{=} \psi_t(z-x). \tag{4.28}$$

But we know that for f (and similarly for g) holds

$$Q_t f(x) \in L^2\left(\frac{dxdt}{t}\right) \text{ and } \int |Q_t f|^2 \frac{dxdt}{t} = c_0 \|f\|_{L^2}^2.$$

This yields the following result.

Lemma 4.36. *If the integral operator with the kernel* (4.28) *is bounded in* $L^2(\mathbb{R}_+^{n+1}, \frac{dxdt}{t})$ *then* T *is bounded in* L^2.

Corollary. *If*

$$\sup_{x,t} \int_{\mathbb{R}_+^{n+1}} |L(x,t;y,s)| \frac{dyds}{s} + \sup_{y,s} \int_{\mathbb{R}_+^{n+1}} |L(x,t;y,s)| \frac{dxdt}{t} < \infty,$$

then T *is bounded in* L^2.

In Chapter 5 we shall see that Lemma 4.36 can in the applications replace the Cotlar-Stein lemma (Lemma 3.6).

5.3. H^1- and BMO-Estimates. Estimates for S-functions and the potentials (4.24) in H^1 and BMO can be obtained in an analogous way as in Sect. 1.7.

Theorem 4.37. (i) $f \in \text{BMO}$ *if and only if the measure*

$$d\mu_f(x,t) = |Q_t f(x)|^2 \frac{dxdt}{t}$$

is a Carleson measure and then $\kappa(\mu_f) \asymp \|f\|_{\text{BMO}}^2$.

(ii) $f \in H^1(\mathbb{R}^n)$ *if and only if* $S(f) \in L^1(\mathbb{R}^n)$.

(iii) $f \in H^1(\mathbb{R}^n)$ *if and only if* f *admits a representation* (4.24) *with* $A(w) \in L^1(\mathbb{R}^n)$.

(iv) $f \in \text{BMO}$ *if and only if* f *admits a representation* (4.24) *such that the measure*

$$|w(x,t)|^2 \frac{dxdt}{t}$$

is a Carleson measure.

Let us indicate the plan of the proof. If $f \in \text{BMO}$ we can show that μ_f is a Carleson measure exactly as in Sect. 1.7. If $f \in H^1$ then the containment $S(f) \in L^1$ follows by Calderón-Zygmund theory: Theorem 3.14 says that every Calderón-Zygmund operator acts from H^1 into L^1 and this includes vector valued operators to which category S(f) belongs.

Now we see from (4.27) that functions in BMO and H^1 admit the desired integral representations. Conversely, if a function f admits the integral representation (4.24) with $A(w) \in L^1$, in particular, if $S(f) \in L^1$ then Theorem 4.6 on atomic decomposition shows that $f \in H^1$. Finally, the sufficiency of the condition for the containment $f \in \mathrm{BMO}$ is derived from the duality theorem exactly as in Sect. 1.7.

The following result due to Coifman, Meyer, and Stein (1983) generalizes Calderón's Theorem 4.7 and illustrates how Theorem 4.37 works in the applications.

Let $n = 1$; we consider a function φ such that $\varphi(x)$, $x\varphi(x)$ and $x^2\varphi(x)$ satisfy the conditions (4.20) and form the corresponding operators $\{P_t\}$, $t > 0$.

Theorem 4.38. *Assume that the functions $u(x,t) = u_t(x)$ and $v(x,t) = v_t(x)$ in $\mathbb{R}^2_+$ are such that*

$$\iint_{\mathbb{R}^2_+} |u(x,t)|^2 \frac{dxdt}{t} < \infty, \quad \int_{-\infty}^{\infty} \sup_t |v(x,t)|^2 dx < \infty.$$

Then the function

$$h(x) = \frac{1}{i}\frac{d}{dx}\int_0^{\infty} (P_t u_t)(x) \cdot (P_t v_t)(x) dt$$

belongs to $H^1(\mathbb{R})$.

In Coifman, Meyer, and Stein (1983) Theorem 4.38 is applied to estimating multiple commutators, which we shall consider in Chapter 5.

Set $w_t(x) = (P_t u_t)(x) \cdot (P_t v_t)(x)$. From the assumptions of Theorem 4.38 it is not hard to deduce that $A(w) \in L^1(\mathbb{R})$, so that by Theorem 4.37 for each admissible ψ holds $\int_0^\infty Q_t w_t \frac{dt}{t} \in H^1$.

Now, if the support of $\hat{\varphi}$ is compact then we can find a ψ such that $Q_t w_t = \frac{1}{i}\frac{d}{dx} w_t$. Indeed, if for example $\operatorname{supp}\hat{\varphi} \subset [-1,1]$ then $\operatorname{supp}\hat{w}_t \subset [-2/t, 2/t]$ and it suffices to take $\psi \in \mathcal{S}(\mathbb{R})$ such that $\hat{\psi}(\xi) = \xi$ for $|\xi| \le 2$. Thus in this case $h \in H^1$.

In the general case we have to represent $\hat{\psi}$ as a sum of functions with compact support, which will be rapidly converging in view of the assumptions on $\hat{\varphi}$ (cf. Coifman, Meyer, and Stein (1983)).

5.4. Paraproducts. A *paraproduct* is a singular integral operator of the form

$$Tf = \int_0^{\infty} Q_t\{(P_t f)(Q_t b)\} \frac{m(t)}{t}\, dt. \tag{4.29}$$

Theorem 4.39. *If $b \in \mathrm{BMO}(\mathbb{R}^n)$, $m \in L^\infty(0,+\infty)$, then T is bounded in L^2 and*

$$\|Tf\|_{L^2(\mathbb{R}^n)} \le c\|bf\|_{\mathrm{BMO}}\|m\|_{L^\infty}\|Tf\|_{L^2(\mathbb{R}^n)}.$$

In fact, if $f \in L^2(\mathbb{R}^n)$ then

$$\int_{\mathbf{R}^n} f(x) \cdot g(x)\, dx = \iint_{\mathbf{R}^{n+1}_+} Q_t g(x) P_t f(x) Q_t b(x) \frac{m(t)}{t}\, dx dt,$$

which in view of Schwarz's inequality does not exceed

$$\|m\|_{L^\infty} \left(\iint_{\mathbf{R}^{n+1}_+} |Q_t g|^2 \frac{dxdt}{t} \right)^{1/2} \left(\iint_{\mathbf{R}^{n+1}_+} |P_t f|^2 |Q_t b|^2 \frac{dxdt}{t} \right)^{1/2} \leq$$
$$\leq c\|m\|_{L^\infty} \|b\|_{\mathrm{BMO}} \|f\|_{L^2} \|g\|_{L^2}.$$

The last inequality follows from the fact that μ_b is a Carleson measure.

An important special case is $m = c_0^{-1}$. Then the operator (4.29) takes the form

$$L_b f = c_0^{-1} \int_0^\infty Q_t\{(P_t f) \cdot (Q_t b)\} \frac{dt}{t}. \tag{4.30}$$

The operator L_b is, as we have established, bounded in L^2 when $b \in \mathrm{BMO}$. It is a singular integral operator, whose kernel is easy to write down. It is a Calderón-Zygmund kernel so that L_b is a Calderón-Zygmund operator. One can verify that the estimates (3.2) and (3.3) hold with the exponent $\alpha = 1$.

The main property of the operator L_b is that

$$L_b(1) = c_0^{-1} \int_0^\infty Q_t^2 b \frac{dt}{t} = b, \tag{4.31}$$

$$L_b^* 1 = 0, \tag{4.32}$$

Formula (4.31) follows from (4.27), while (4.32) follows from (4.21), using that $Q_t^* 1 \equiv 0$.

Thus, given a function $b \in \mathrm{BMO}$ one can construct a Calderón-Zygmund operator L_b such that $L_b 1 = b$ and $L_b^* 1 = 0$. If b_1 and b_2 are two BMO functions then the operator $L_{12} = L_{b_1} - L_{b_2}^*$ is such that $L_{12}(1) = b_1$ and $L_{12}^*(1) = b_2$.

This construction will play a decisive rôle in the proof of the David-Journé theorem in Chapter 5.

In the general theorem of Coifman and Meyer (1978) a superfluous (compared to (4.29)) convolution Q_t is removed:

Theorem 4.40. *Assume that the kernels φ and ψ are such that $\hat{\varphi}(\xi)$ and $\hat{\psi}(\xi)$ coincide for $|\xi|$ sufficiently large with functions in the Schwartz space $S(\mathbb{R}^n)$ and, moreover, that for every multi-index α*

$$|\partial^\alpha \hat{\psi}(\xi)| \leq c_\alpha |\xi|^{1-|\alpha|}.$$

Then the operator

$$Tf = \int_0^\infty P_t f \cdot Q_t b \frac{m(t)}{t}\, dt$$

is bounded in L^2 *for any choice of* $b \in \mathrm{BMO}(\mathbb{R}^n)$ *and* $m \in L^\infty(0,\infty)$, *and*

$$\|Tf\|_{L^2} \le c\|m\|_{L^\infty}\|b\|_{\mathrm{BMO}}\|f\|_{L^2}.$$

In particular, if φ is the Poisson kernel, $n = 1$, and $\psi_t = -t\frac{\partial\varphi_t}{\partial t}$, the functions f and b being analytic in the upper halfplane $\mathbb{C}_+$, then

$$Tf(x) = \int_{-\infty}^{x} b'(s)f(s)ds$$

and so from Theorem 4.40 we obtain the following result due to Pommerenke (1978):

Corollary. *If* $f \in \mathrm{BMO}A(\mathbb{C}_+)$, $f \in H^2(\mathbb{C}_+)$, *then the function*

$$L(z) = \int_{-\infty}^{z} b'(s)f(s)ds$$

belongs to $H^1(\mathbb{C}_+)$.

Compare this corollary with Calderón's Theorem 4.7.

§6. The Coifman Construction

Until recently all variants of the Littlewood-Paley theory were based either on Fourier analysis or on Green's formula. Such a theory was not sufficiently flexible and, in particular, it did not carry over to spaces of homogeneous type (Coifman and Weiss (1977)).

Not long ago Coifman gave the first purely geometric variant of the theory. Coifman's construction is unpublished, and we shall set it forth following the papers David, Journé, and Semmes (1985), (1986).

In Sect. 6.1 this construction is described and the main estimates of Littlewood-Paley type are formulated. In Sect. 6.2 we discuss the scheme of the proof of these estimates, while in Sect. 6.3 we give the construction of the paraproducts in the Coifman construction.

6.1. Formulation of the Results. Let us consider in $\mathbb{R}^n$ a "*quasi-unity*", that is, a sequence of kernels $\{s_k(x,y)\}_{-\infty}^{\infty}$, $x, y \in \mathbb{R}^n$, such that

(i) $s_k(x,y) = 0$ for $|x-y| > 2^{-k}$,

(ii) $|s_k(x,y)| \le c \cdot 2^{kn}$,

(iii) $s_k(x,y) = s_k(y,x)$,

(iv) $\int_{\mathbb{R}^n} s_k(x,y)\,dy = 1$,

(v) $|s_k(x,y) - s_k(x',y)| \le c \cdot 2^{k(n+\alpha)}|x-x'|^\alpha$, $\alpha > 0$.

The kernels $\{s_k\}$ generate a sequence of smoothing operators in the scale 2^{-k}:

$$S_k f(x) = \int_{\mathbb{R}^n} s_k(x,y)f(y)\,dy, \quad k = 0, \pm 1, \pm 2, \dots .$$

It is clear that $\lim_{k\to\infty} S_k f = f$, $\lim_{k\to-\infty} S_k f = 0$, for example in L^2. Set

$$D_k = S_k - S_{k-1}.$$

The operators D_k are selfadjoint in L^2 and

$$\|D_j D_k\| \le c \cdot 2^{-\alpha|j-k|}. \tag{4.33}$$

Lemma 4.41. (i) *For any function $f \in L^2$*

$$\sum_k \|D_k f\|_{L^2}^2 \le c\|f\|_{L^2}^2.$$

(ii) *If $b \in \mathrm{BMO}(\mathbb{R}^n)$ then for each $x \in \mathbb{R}^n$ and $r > 0$*

$$\sum_{2^{-k}<r} \int_{|y-x|<r} |D_k b(y)|^2\, dy \le c\|b\|_{\mathrm{BMO}}^2\, r^n.$$

Remark. Of course, (ii) means that μ_b, with

$$d\mu_b(x,t) = \sum_k |D_k b(x)|^2\, dx \otimes \delta_k(t),$$

where $\delta_k(t)$ denotes a unit mass at $t = 2^{-k}$, is a Carleson measure in $\mathbb{R}_+^{n+1}$.

Next, set

$$D_j^n = \sum_{|j-k|\le n} D_k$$

and

$$T_n = \sum_{j=-\infty}^{\infty} D_j D_j^n = \sum_{|j-k|\le n} D_j D_k, \quad n = 1, 2, \dots$$

(this series is weakly convergent in L^2). Clearly T_n is a selfadjoint operator in L^2.

Theorem 4.42. *$T_n \to 1$ in the L^2 operator norm.*

Corollary 1. *T_n is invertible in L^2 for n sufficiently large.*

Corollary 2. *T_n is bounded and invertible in* $\mathrm{BMO}(\mathbb{R}^n)$ *and* $H^1(\mathbb{R}^n)$ *for n sufficiently large.*

Corollary 3. *T_n is bounded and invertible in L^p, $1 < p < \infty$, for n sufficiently large.*

In view of Theorem 4.42 we can instead of $\|f\|$ estimate the norm of

$$T_n f = \sum_j D_j D_j^n f = \sum_k D_k^n D_k f.$$

This identity is analogous to the integral representations (4.6) or (4.27). For example, if $g \in L^2$ then

$$
\begin{aligned}
|\langle T_n f, g\rangle| = \Big|\sum_j \langle D_j f, D_j^n g\rangle\Big| &\le \sum_j \|D_j f\| \cdot \|D_j^n g\| \\
&\le \Big(\sum \|D_j f\|^2\Big)^{1/2} (2n+1) \Big(\sum \|D_j g\|^2\Big)^{1/2} \\
&\le (2n+1)\, c \cdot \|g\|_{L^2} \cdot \Big(\sum \|D_j f\|^2\Big)^{1/2},
\end{aligned}
$$

and we have the converse estimate to Lemma 4.41 (i):

$$\|f\|_{L^2}^2 \le c \sum \|D_j f\|_{L^2}^2.$$

Furthermore, let T be the linear operator whose L^2 boundedness we wish to investigate.

It is sufficient to establish the boundedness of the operator $T_n T T_n$. But $T_n T T_n = \sum_{j,k} D_j^n D_j T D_k D_k^n$, so that for $f, g \in L^2$

$$|\langle T_n T T_n f, g\rangle| \le \sum_{j,k} \|D_j^n f\|_{L^2}\, \|D_j T D_k\|\, \|D_k^n g\|_{L^2}.$$

By Lemma 4.41 we know that $\sum \|D_j^n f\|_{L^2}^2 \le c(2n+1)^2 \|f\|_{L^2}^2$, whence we obtain the following result.

Lemma 4.43. *If the matrix*

$$a_{jk} = \|D_j T D_k\|$$

defines a bounded linear operator in $l^2(\mathbb{Z})$, then T is bounded in $L^2(\mathbb{R}^n)$.

In particular, the hypothesis of Lemma 4.43 is satisfied if

$$\sum_j \|D_j T D_k\| + \sum_k \|D_j T D_k\| \le c < \infty.$$

Of course, this is the complete analogue of Lemma 4.36, while the invertiblity of T_n corresponds to the representation (4.27).

Theorem 4.42 does not give us the analogue of the full Littlewood-Paley theory but only of the basic integral representation (4.6) on (4.27). It is possible to give an analogue also of other portions of the theory. The analogue of the potential (4.7) is

$$h = \sum_j D_j h_j. \tag{4.34}$$

For example, if $g \in L^2$ then

$$\langle h, g\rangle = \sum_j \langle D_j h_j, g\rangle = \sum_j \langle h_j, D_j g\rangle,$$

which gives the estimate

$$\|h\|_{L^2}^2 \le c \sum \|h_j\|_{L^2}^2 .$$

Thus,

$$g(f)(x) = \left(\sum_j |D_j f(x)|^2 \right)^{1/2}, \quad x \in \mathbb{R}^n,$$

ought to be viewed as the analogue of the Littlewood-Paley g-function, while

$$S(f)(x) = \left(\sum_j 2^j \int_{|y-x|<2^{-j}} |D_j f(y)|^2 \, dy \right)^{1/2}$$

is the analogue of the S-function. In particular, we have

Theorem 4.34. *If* $1 < p < \infty$ *then*

$$\|S(f)\|_{L^p(\mathbb{R}^n)} \asymp \|f\|_{L^p(\mathbb{R}^n)} .$$

for $f \in L^p(\mathbb{R}^n)$.

6.2. Scheme of Proof. The estimate (4.33) is a consequence of direct estimates for the kernel of the operator $D_j D_k$. If

$$v_j(x,y) = s_j(x,y) - s_{j-1}(x,y),$$

then $\int v_j dy = 0$ and the kernel $D_j D_k$ for $j > k$ equals

$$\int_{\mathbb{R}^n} v_j(x,z) v_k(z,y) \, dz = \int_{\mathbb{R}^n} v_j(x,z) [v_k(z,y) - v_k(x,y)] \, dz.$$

The last integral can immediately be estimated in view of the conditions (i), (ii) and (v) on the kernels $\{s_j\}$. Lemma 4.41 (i) is a simple consequence of (4.33), because

$$\begin{aligned} \left(\sum \|D_k f\|_{L^2}^2 \right)^{1/2} &= \sup \left\{ \sum \langle D_k f, g_k \rangle, \sum \|g_k\|_{L^2}^2 \le 1 \right\} \\ &\le \|f\| \sup \left\{ \|\sum D_k g_k\|_{L^2}, \sum \|g_k\|_{L^2}^2 \le 1 \right\}, \end{aligned}$$

whereas

$$\left\| \sum D_k g_k \right\|_{L^2}^2 = \sum_{j,k} \langle D_j D_k g_k, g_j \rangle \le \sum_{j,k} \|D_j D_k\| \, \|g_j\| \, \|g_k\| .$$

Lemma 4.41 (ii) folows from (i) in the same way as in Sect. 1.7 the BMO estimate was derived from the L^2-estimates.

Let us now turn to Theorem 4.42. Indeed, we obtain the more precise estimate

$$\|T_n - \mathbf{1}\| \le c \cdot 2^{-\frac{\alpha}{2}n}, \quad n \to \infty. \tag{4.35}$$

Without loss of generality we may assume that $\sum \|D_j\| < \infty$ and still get the estimate (4.35) with a value of c, which is independent of this assumption.

Lemma 4.46. $\|T_n(\mathbf{1} - T_m)T_n\| \le c \cdot 2^{-\frac{\alpha}{2}m}(1 + n^2)$.

The reasoning preceeding Lemma 4.43 shows that it suffices to estimate the $l^2(\mathbb{Z})$ norm of the matrix

$$a_{jk} = \|D_j(\mathbf{1} - T_m)D_k\| \le \sum_{|s-t|>m} \|D_j D_s D_t D_k\|.$$

But

$$\begin{aligned}\|D_j D_s D_t D_k\| &= \|(D_j D_s)(D_t D_k)\| = \|D_j(D_s D_t)D_k\| \\ &\le \min\{c \cdot 2^{-\alpha(|s-j|+|t-k|)}, c \cdot 2^{-\alpha|s-t|}\},\end{aligned}$$

whence

$$\sum_k a_{jk} \le c \cdot 2^{-\frac{\alpha}{2}m}, \quad \sum_j a_{jk} \le c \cdot 2^{-\frac{\alpha}{2}m}.$$

Next, let n_0 be the least value of n for which $\|T_{n_0} - \mathbf{1}\| \le \frac{1}{2}$. Then by Lemma 4.46

$$\frac{1}{2} \le \|T_{n_0-1} - \mathbf{1}\| \le \|T_{n_0}^{-1}\| \|T_{n_0}(T_{n_0-1} - \mathbf{1})T_{n_0}\| \|T_{n_0}^{-1}\| \le c(n_0^2 + 1)2^{-\frac{\alpha}{2}n_0},$$

that is, $n_0 \le c_1$, and applying Lemma 4.46 anew we find

$$\|\mathbf{1} - T_n\| \le 4\|T_{n_0}(\mathbf{1} - T_n)T_{n_0}\| \le c_2 \cdot 2^{-\frac{\alpha}{2}n},$$

which concides with (4.35).

Corollary 1 and 2 follow from Theorem 4.42, because T_n, as is readily seen, is a Calderón-Zygmund operator. Finally, Corollary 3 follows from Calderón-Zygmund theory, along with the fact that $(\mathbf{1} - T_n)(1) = 0$ (cf. Sect. 2 in Chapter 5).

6.3. Paraproducts. Let $b \in \mathrm{BMO}(\mathbb{R}^n)$. The analogue of the *paraproduct* L_b (4.30) in the Coifman construction is the operator

$$Lf = \sum_k D_k\{(D_k^n \gamma) \cdot S_k f\},$$

where $\gamma = T_n^{-1}b$ and n is sufficiently large. It is clear that

$$L(1) = \sum D_k D_k^n \gamma = T_n \gamma = b$$

and that

$$L^*(1) = 0.$$

It is not hard to check that the kernel of L is a Calderón-Zygmund kernel. Let us show that L is bounded in L^2. We have for $f, g \in L^2$

$$\begin{aligned} |\langle Lf, g\rangle| &\leq \sum \|D_k g\| \, \|(D_k^n \gamma) \cdot S_k f\| \\ &\leq \left(\sum \|D_k g\|^2\right)^{1/2} \left(\sum \|(D_k^n \gamma) \cdot S_k f\|^2\right)^{1/2} \\ &\leq c\|g\|_{L^2} \left(\iint_{\mathbb{R}_+^{n+1}} |F(x,t)|^2 d\mu_\gamma(x,t)\right)^{1/2}. \end{aligned}$$

Here μ_γ is the measure described in the remark to Lemma 4.41 (ii), while $F(x,t) = S_k f(x)$ for $t = 2^{-k}$. As the nontangential maximal function F does not exceed MF, the desired estimate

$$|\langle Lf, g\rangle| \leq c\|f\|_{L^2} \|g\|_{L^2}$$

follows from Calderón-Zygmund theory.

Thus, L is a Calderón-Zygmund operator.

§7. Fourier Multipliers and the Dyadic Expansion

7.1. Application of g Functions. Let m be a bounded measurable function in $\mathbb{R}^n$. The operator $T_m : L^2(\mathbb{R}^n) \to L^2(\mathbb{R}^n)$,

$$(T_m f)\hat{\ }(\xi) = m(\xi)\hat{f}(\xi), \quad \xi \in \mathbb{R}^n,$$

clearly is bounded, and $\|T_m\| = \|m\|_{L^\infty}$. If it is bounded in the L^p norm,

$$\|T_m f\|_{L^p} \leq c\|f\|_{L^p}, \quad f \in L^p \cap L^2,$$

we say that m is a *multiplier* in L^p.

It is not hard to see that the spaces L^p and $L^{p'}$ admit the same multipliers, and that the multipliers in L^1 are precisely the Fourier transforms of finite Borel measures.

Littlewood-Paley theory gives sufficient conditions for a function m to be a multiplier in L^p for all p, $1 < p < \infty$. These conditions impose additional smoothness of m in $\mathbb{R}^n \setminus \{0\}$. This part of the theory is very well explained in Stein's book (1970a). Therefore we give here just the formulations of the main results with short comments. For further information we refer to Larsen (1971).

Our first result is the *multiplier theorem of Marcinkiewicz-Mikhlin-Hörmander.*

Theorem 4.47. *Let* $k = [\frac{n}{2}] + 1$, $m \in C^k(\mathbb{R}^n \backslash \{0\})$ *and assume that for each multi-index* α *with* $|\alpha| \leq k$ *holds*

$$\sup_{0<R<\infty} R^{2|\alpha|-n} \int_{R<|\xi|<2R} |\partial^\alpha m(\xi)|^2 d\xi < \infty.$$

Then m *is a multiplier on* L^p *for all* p, $1 < p < \infty$.

Corollary. *If* $|\partial^\alpha m(\xi)| \leq c_\alpha |\xi|^{-|\alpha|}$, $|\alpha| \leq k$, *the* m *is a multiplier on* L^p *for* p, $1 < p < \infty$.

Theorem 4.47 is applicable to, for instance, the function $|\xi|^{i\tau}$ with $\tau \in \mathbb{R}$, and further to any function in $\mathbb{R}^n$ that is homogeneous of degree 0 and of class $C^k(\mathbb{R}^n \backslash \{0\})$.

The proof of Theorem 4.47 is based on the following estimate:

$$g_0(T_m f)(x) \leq c g_{\frac{2k}{n}}(f)(x), \quad x \in \mathbb{R}^n.$$

It is clear that the theorem follows from this along with the results of Sect. 4. The main estimate is proved in the immediate way: the Poisson integral of $T_m f$ is represented as a convolution of f with a certain kernel, while the hypothesis on m allows us to estimate this kernel (cf. Stein (1970a), Chapter IV).

7.2. The Dyadic Expansion. Further progress in the theory of multipliers is connected with the famous *Littlewood-Paley theorem on the dyadic expansion.*

Let us divide $\mathbb{R}$ into dyadic intervals

$$I_j^\pm = [\pm 2^j, \pm 2^{j+1}], \quad -\infty < j < \infty.$$

Forming all possible products $Q_j = I_{j_1} \times I_{j_2} \times \dots I_{j_n}$ we get a decomposition of $\mathbb{R}^n$ into dyadic parallelotopes. Let S_j be the operator of forming a partial sum of the Fourier transform:

$$(S_j f)\hat{}\,(\xi) = \hat{f}(\xi)\chi_{Q_j}(\xi),$$

where χ_{Q_j} is the characteristic function of Q_j. Then $f = \sum_j S_j f$ and, the term being orthogonal, we have

$$\|f\|_{L^2} = \left\| \left(\sum_j |S_j f|^2 \right)^{1/2} \right\|_{L^2}.$$

Theorem 4.48. *For any p,* $1 < p < \infty$,

$$\|f\|_{L^p} \asymp \Big\| \Big(\sum_j |S_j f|^2\Big)^{1/2} \Big\|_{L^p}.$$

Theorem 4.48 is an anisotropic statement, because the Q_j are not cubes, and their dimensions along various coordinate axes are not related. Therefore, it is no surprise that the proof of the n-dimensional case is obtained by superposition of the corresponding one-dimensional estimates. In the one-dimensional case Theorem 4.48 is reduced to Theorem 4.47 with the aid of the following device.

Consider a sequence of functions $\varphi_j \in \mathcal{D}(\mathbb{R})$ such that $0 \le \varphi_j \le 1$, $\varphi_j(\xi) = 1$ for $\xi \in I_j$ and $\varphi_j(\xi) = 0$ for $\rho(\xi, I_j) > 2^{j-2}$ and, finally, $|\varphi_j'(\xi)| \le 2^{2-j}$. Let us replace S_j by the multiplier transform $T_j = T_{\varphi_j}$.

Lemma 4.49. $\|(\sum_j |T_j f|^2)^{1/2}\|_{L^p} \le c_p \|f\|_{L^p}$, $1 < p < \infty$.

Indeed, the multiplier

$$m_t(\xi) = \sum_j r_j(t)\varphi_j(\xi),$$

where r_j are the Rademacher functions (Sect. 6 of Chapter 1), satisfies uniformly the hypotheses of Theorem 4.47. Therefore,

$$\int \Big|\sum_j r_j(t) T_j f(t)\Big|^p dx \le c_p \int |f|^p.$$

It suffices to integrate this inequality with respect to t.

In order to replace in Lemma 4.49 T_j by S_j we remark that $S_j = S_j T_j$, as $\chi_{I_j} = \chi_{I_j}\varphi_j$, and apply the following lemma.

Lemma 4.50. *If* $\{f_j\} \in L^p$ *then*

$$\Big\| \Big(\sum_j |S_j f_j|^2\Big)^{1/2} \Big\|_{L^p} \le c_p \Big\| \Big(\sum_j |f_j|^2\Big)^{1/2} \Big\|_{L^p}, \quad 1 < p < \infty.$$

Indeed, in view of Sect. 2.2 of Chapter 2, the partial sum operator S_j can be expressed in terms of the Hilbert transform, and Lemma 4.50 follows from a vector valued estimate for the Hilbert transform (Sect. 7.2 of Chapter 3).

The converse estimate $\|f\|_{L^p} \le c_p \|(\sum_j |S_j f_j|^2)^{1/2}\|_{L^p}$ follows by duality considerations.

7.3. Application of the Dyadic Expansion. Let us apply Theorem 4.48 to estimates for multipliers, for simplicity restricting ourselves to the case

$n = 1$. Let m be a function on $\mathbb{R}$. In order to estimate the norm of $T_m f$ it is, in view of Theorem 4.48, sufficient to estimate the norm of the expression $(\sum_j |T_m S_j f|^2)^{1/2}$. Let $I_j = [2^j, 2^{j+1})$. If $\xi \in I_j$ we have

$$m(\xi) = m(2^j) + \int_{2^j}^{\xi} m'(t)dt.$$

Let S_t be the partial sum operator corresponding to the interval $[2^j, t)$. Then

$$\begin{aligned} T_m S_j &= m(2^j)S_j + \int_{2^j}^{2^{j+1}} m'(t)(S_j - S_t)dt \\ &= m(2^{j+1})S_j - \int_{2^j}^{2^{j+1}} m'(t)S_t dt. \end{aligned}$$

Therefore

$$|T_m S_j f|^2 \le |m(2^{j+1})|^2 |S_j f|^2 + \left(\int_{2^j}^{2^{j+1}} |m'(t)||S_j S_t f|\, dt \right)^2 .$$

Thus we obtain the following result.

Theorem 4.51. *If m is in $L^\infty(\mathbb{R}^n)$ and if for each dyadic interval $I_j^\pm$ we have*

$$\int_{I_j^\pm} |dm(\xi)| < B < \infty,$$

then m is a multiplier in L^p for $1 < p < \infty$.

Iterating this result we can prove the n-dimensional analogue of Theorem 4.51. Let us assume that the function $m(\xi)$ is defined on the set of all points $(\xi_1, \ldots, \xi_n) \in \mathbb{R}^n$ such that all $\xi_i \neq 0$ and admits on this set continuous derivatives $\partial^\alpha m(\xi)$, where $\alpha = (\alpha_1, \ldots, \alpha_n)$ and each index α_i takes only the values 0 and 1.

Theorem 4.52. *If for $0 \le k < n$ and each dyadic parallelotope $Q \subset \mathbb{R}^k$*

$$\sup_{\xi_{k+1}, \ldots, \xi_n} \int_Q \left| \frac{\partial^k m(\xi)}{\partial \xi_1 \ldots \partial \xi_k} \right| d\xi_1 \ldots \xi_k < B < \infty,$$

and if the same estimate holds also for an arbitrary permutation of the coordinates $(\xi_1, \ldots, \xi_n)$, then m is a multiplier in $L^p(\mathbb{R}^n)$, $1 < p < \infty$.

Corollary. *If $|\partial^\alpha m(\xi)| \le c_\alpha |\xi_1|^{-\alpha_1} \ldots |\xi_n|^{-\alpha_n}$, where all α_i are equal to 0 or 1, then m is a multiplier in $L^p(\mathbb{R}^n)$, $1 < p < \infty$.*

It is clear that Theorem 4.52, in contrast to Theorem 4.47, depends on the choice of coordinates in $\mathbb{R}^n$. It is applicable to functions such as $\dfrac{\xi_1}{\xi_1 + i(\xi_2^2 + \cdots + \xi_n^2)^{1/2}}$ and $\dfrac{\xi_1^{\alpha_1}\xi_2^{\alpha_2}\cdots\xi_n^{\alpha_n}}{(1+|\xi|^2)^{|\alpha|/2}}$ etc.

7.4. Weighted and Vectorial Analogues. From the proof of Theorem 4.47 it is clear that it remains in force also for weighted L^p-norms:

$$\|T_m f\|_{L^p_w} \le c\|f\|_{L^p_w}$$

provided $\frac{n}{k} < p < \infty$ and the weight function w satisfies simultaneously the conditions (A_p) and $(A_{\frac{pk}{n}})$.

Anisotropic versions of Theorem 4.48 and 4.52 cause more trouble. It is necessary to replace condition (A_p) by stronger conditions. Let $\mathcal{R}$ be the family of all rectangular parallelotopes in $\mathbb{R}^n$, whose axes are parallel to the coordinate axes. We say that w satisfies the *condition* (A'_p), $1 < p < \infty$, if

$$\sup_{R\subset\mathcal{R}} \left(\frac{1}{|R|}\int_R w\right)\left(\frac{1}{|R|}\int_R w^{-\frac{1}{p-1}}\right)^{p-1} < \infty.$$

This means, in particular, that condition (A_p) is satisfied in each variable individually and then uniformly in all the remaining variables.

Kurtz (1980) has obtained the following weighted analogues of Theorem 4.48 and Theorem 4.52.

Theorem 4.48′. *If* $1 < p < \infty$ *and* w *satisfies condition* (A'_p) *then*

$$\|f\|_{L^p_w} \asymp \left\|\left(\sum |S_j f|^2\right)^{1/2}\right\|_{L^p_w}.$$

Theorem 4.52′. *If in the hypothesis of Theorem 4.52 the weight* w *satisfies* (A'_p) *then* m *is a multiplier in* $L^p_w(\mathbb{R}^n)$.

In Triebel's book (1978) one can find an extension of Theorem 4.47 to the vectorial and matricial cases. Let us state his result.

Theorem 4.53. *Let* $\{m_j\}$ *be a sequence of functions in* $C^k(\mathbb{R}^n\backslash\{0\})$, $k = [n/2]+1$, *which for* $|\alpha| \le k$ *satisfies*

$$\sup_{0<R<\infty} R^{2|\alpha|-n}\int_{R<|\xi|<2R}\left\{\sum|\partial^\alpha m_j(\xi)|^2\right\}d\xi < \infty.$$

Then for any p *and* r, $1 < p < \infty$, $1 < r < \infty$, *and any sequence* $\{f_j\}$ *of functions in* $L^p(\mathbb{R}^n)$ *holds*

$$\left\|\left(\sum |T_{m_j} f_j|^r\right)^{1/r}\right\|_{L^p(\mathbb{R}^n)} \le c\left\|\left(\sum |f_j|^r\right)^{1/r}\right\|_{L^p(\mathbb{R}^n)}.$$

In this section we have stated some conditions that are sufficient for a function m to be a multiplier in L^p. They are far from being optimal and, for instance, they do not discriminate between various values of p between 1 and ∞. Sufficiently general criteria that are free from any deficiencies are so far not known. Let us now give some characteristic examples.

1) A convex polytope in $\mathbb{R}^n$ is the intersection of finitely many halfspaces. Therefore its characteristic function is a multiplier in all L^p, $1 < p < \infty$.

2) But the characteristic function of a ball in $\mathbb{R}^n$ is not a multiplier in any L^p except L^2! This remarkable result is due to Fefferman (1972).

3) Let $\varphi_0 \in C^\infty(\mathbb{R}^n)$, $\varphi_0(\xi) = 0$ for $|\xi| \leq 1$ and $\varphi_0(\xi) = 1$ for $|\xi| \geq 2$. Then the function

$$m(\xi) = \varphi_0(\xi) e^{i|\xi|^\alpha}/|\xi|^\beta, \quad \alpha > 0, \beta > 0$$

is a multiplier in $L^p(\mathbb{R}^n)$ provided

$$\left|\frac{1}{2} - \frac{1}{p}\right| < \frac{\beta}{\alpha n}$$

and only for these values of p.

4) The exact values of α for which the function

$$m(\xi) = \begin{cases} (1 - |\xi|^2)^\alpha, & |\xi| < 1, \\ 0, & |\xi| > 1, \end{cases}$$

is a multiplier in $L^p(\mathbb{R}^n)$ are not known.

Further information on multipliers can be found in the books Larsen (1971), Stein (1970a) and Triebel (1978) and likewise in the article by Alimov et al. in this volume.

§8. Supplements

8.1. Martingale Inequalities. There is an analogue of Littlewood-Paley theory for martingales in probability theory. Here we list briefly the main inequalities. A more detailed discussion may be found in Gikhman and Skorokhod (1982), Burkholder (1979a), (1979b), Garcia (1973), Gundi (1980).

Let $(\Omega, \mathcal{F}, P)$ be a probability space and let $\{\mathcal{F}_n\}_0^\infty$ be a sequence of σ-algebras in it, $\mathcal{F}_n \subset \mathcal{F}$, $\mathcal{F}_n \subset \mathcal{F}_{n+1}$. For each $\mathcal{F}$ measurable function $f \in L^1(\Omega)$ we can define its *conditional expectation* $\mathcal{E}(f|\mathcal{F}_n)$, which is a $\mathcal{F}_n$ measurable function ξ_n such that

$$\int_A f dP = \int_A \xi_n dP, \quad A \in \mathcal{F}_n.$$

An integer valued random variable τ on Ω is termed a *stopping time* if for any n the set $\{\omega : \tau(\omega) \leq n\}$ is $\mathcal{F}_n$ measurable.

A sequence of random variables $\{\xi_n\}_0^\infty$ is called a *martingale* if

(i) the function ξ_n is $\mathcal{F}_n$ measurable for each n;

(ii) $\mathcal{E}|\xi_n| < \infty$ for all n;

(iii) $\xi_m = \mathcal{E}(\xi_n|\mathcal{F}_m)$ for $m < n$.

For example, if $\gamma(t)$ is Brownian motion in the plane and u is a harmonic function, then $\{u(\gamma(n))\}_0^\infty$ is a martingale. On the other hand, the collection of conditional expectations $\{\mathcal{E}(f|\mathcal{F}_n)\}$ of any given function f is a martingale. Note that we consider here only martingales in discrete time, this to avoid technical complications.

Set

$$\|\xi\|_p = \sup_n(\mathcal{E}|\xi_n|^p)^{1/p}, \quad 1 < p < \infty.$$

A sequence of random variables $\{a_n\}_0^\infty$ is said to be *predictable* if for all n the random variable a_n is $\mathcal{F}_{n-1}$ measurable. With each martingale $\{\xi_n\}$ we may associate the *maximal function*

$$\xi^* = \sup_n |\xi_n|$$

and the *Littlewood-Paley function*

$$g(\xi) = |\xi_0| + \left(\sum_{n=0}^{\infty} |\xi_{n+1} - \xi_n|^2\right)^{1/2}$$

These are random variables on Ω.

Lastly, let $\{a_n\}$ be an arbitrary sequence of random variables. Then we can define the martingale transform $\zeta = a \circ \xi$:

$$\zeta_n = a_0\xi_0 + a_1(\xi_1 - \xi_0) + \dots a_n(\xi_n - \xi_{n-1}).$$

One can show that if the sequence $\{a_n\}$ is predictable then $\{\zeta_n\}$ is again a martingale. If τ is a stopping time then $\{\zeta_{\min(n,\tau)}\}$ too is a martingale.

Theorem 4.54. *If ξ is a martingale then*
(i) $P\{\xi^* > \lambda\} \le 3/\lambda\|\xi\|_1$, $\lambda > 0$;
(ii) $(\mathcal{E}|\xi^*|^p)^{1/p} \le c_p\|\xi\|_p$, $1 < p < \infty$.

Theorem 4.55. *If ξ is a martingale and $\|\xi\|_1 < \infty$ then $\xi_\infty = \lim \xi_n$ exists almost surely and $\xi_n = \mathcal{E}(\xi_\infty|\mathcal{F}_n)$. Moreover, if $\|\xi\|_p < \infty$, $1 < p < \infty$, then* $\mathcal{E}|\xi_n - \xi_\infty|^p \to 0$.

These theorems are due to Doob.

Theorem 4.56. *Let ξ be a martingale and a a predictable sequence with $|a_n| \le 1$. Then*
(i) $P\{(a \circ \xi)^* > \lambda\} \le 2/\lambda\|\xi\|_1$, $\lambda > 0$;
(ii) $\|a \circ \xi\|_p \le c_p\|\xi\|_p$, $1 < p < \infty$.

If we take instead of $\{a_n\}$ the deterministic sequence $a_n = r_n(t)$, where r_n are the Rademacher functions (Sect. 6 of Chapter 1) and integrate with respect to t, we readily find

Theorem 4.57. *If ξ is a martingale then*

$$(\mathcal{E}|\xi_\infty|^p)^{1/p} \le c_p(\mathcal{E}(g(\xi))^p)^{1/p} \le c_p\|\xi\|_p, \quad 1 < p < \infty.$$

Theorem 4.57 clearly is the analogue of the Littlewood-Paley estimates for martingales. The last two theorems were first proved by Burkholder, and Gundy (1979a), (1979b) and Gundy (1980).

8.2. Wavelets. In recent work of Morlet, Meyer and their associates (see Battle and Federbush (1982), Daubechies, Grossmann, and Meyer (1986), Lemarié, and Meyer (1986), Meyer (1987)) there is constructed a remarkable class of orthogonal systems in $L^2(\mathbb{R}^n)$. We restrict ourselves to formulating some results for $n = 1$.

Let $m > 0$ be an integer. There exists a function ψ on $\mathbb{R}$ enjoying the following properties:

(i) $\psi \in C^{m-1}(\mathbb{R})$ and $\psi^{(m)} \in L^\infty$.

(ii) $\psi^{(m)}(x)$, $k = 0, 1, \dots, m$, is rapidly decreasing as $x \to \infty$.

(iii) $\int_{-\infty}^{\infty} x^k\psi(x)dx = 0$, $k = 0, 1, \dots, m$.

(iv) The functions $\psi^{jk}(x) = 2^{j/2}\psi(2^jx - k)$, $j \in \mathbb{Z}$, $k \in \mathbb{Z}$, form an orthonormal basis in $L^2(\mathbb{R})$.

For $m = 0$ the conditions (i)–(iv) are fulfilled by the basis functions of the Haar system:

$$\psi(x) = \begin{cases} 1, & 0 \le x < 1/2 \\ -1, & 1/2 \le x \le 1/2 \\ 0, & x \notin [0, 1] \end{cases}$$

For $m > 0$ or for $m = \infty$ the existence of such functions ψ is not obvious, but in Meyer (1987) an explicit construction of them is given. If $m = 1$ we have a particularly simple picture of ψ: it is a piecewise linear function with knots at integer points (that is, a linear spline). In practice the rate of the "rapid decrease" in condition (ii) depends on m: for finite m one can achieve exponential decrease but for $m = \infty$ it is known that ψ drops off faster than any power, but not exponentially.

Let $\mathcal{G}$ be the set of all *dyadic intervals* on $\mathbb{R}$, that is, intervals of the form

$$I = I_{jk} = [k \cdot 2^{-j}, (k+1) \cdot 2^{-j}].$$

Set

$$\psi_I(x) = \psi_{jk}(x) = 2^{j/2}\psi(2^jx - k).$$

The basis function ψ_I is localized close to the interval I, so that, for example, $\|\psi_I\chi_{\mathbf{R}\setminus\lambda I}\|_{L^2}$ falls off rapidly with growing λ.

The functions ψ_I are called *wavelets* (French: *ondelettes*) and the orthogonal expansion

$$f = \sum_{I\in\mathcal{G}} \alpha(I)\psi_I, \quad \alpha(I) = \langle f, \psi_I\rangle, \tag{4.36}$$

of a function $f \in L^2$ is called the wavelet expansion.

Of course, $f \in L^2$ if and only if $\sum |\alpha(I)|^2 < +\infty$. But it turns out that that membership of a function f in an arbitrary space L^p, $1 < p < \infty$, or H^p, $p > 0$, in BMO or in a space of smooth functions (for example, Λ^s) also can be expressed by conditions imposed just on the moduli of the coefficients $\alpha(I)$!

In the case L^p one introduces to this end the *Luzin function*

$$S(f)(x) = \left(\sum_{I \ni x} |\alpha(I)|^2 |I|^{-1} \right)^{1/2}, \quad x \in \mathbb{R}.$$

Theorem 4.58. (i) *$f \in L^p$ if and only if $S(f) \in L^p$, $1 < p < \infty$.*
(ii) *$f \in H^1(\mathbb{R})$ if and only if $S(f) \in L^1$.*

While Theorem 4.58 is true for any $m > 0$, a characterization of H^p, $0 < p < 1$, is possible only for sufficiently large m.

Corollary. *Wavelets give an unconditional basis in L^p, $1 < p < \infty$, and in H^1.*

Remark. The fact that wavelets give an unconditional basis can be proved in a very elegant way without using Littlewood-Paley theory (Meyer (1987)). It is sufficient to establish the inequality

$$\sup_{|\lambda_I| \le 1} \left\| \sum_I \lambda_I \alpha(I) \psi_I \right\| \le c \left\| \sum_I \alpha(I) \psi_I \right\|.$$

Consider the operator

$$T : \sum_I \alpha(I) \psi_I \to \sum_I \lambda_I \alpha(I) \psi_I.$$

It is a Calderón-Zygmund operator with the kernel

$$K(x, y) = \sum_I \lambda_I \psi_I(x) \psi_I(y).$$

That it is bounded in L^2 is obvious and that we have a uniform estimate in λ_I of the kernel follows from the conditions (i)–(iv).

Thus, it is uniformly (in λ) bounded in all spaces L^p, $1 < p < \infty$, and, as $T^*(1) = 0$, likewise in H^1 (cf. Sect. 4.3 of Chapter 5).

Theorem 4.59. $f \in \mathrm{BMO}$ *if and only if*

$$\sum_{I \subset J} |\alpha(I)|^2 \le \kappa |J|, \quad J \in \mathcal{G}.$$

We see that with the expansion (4.36) there is connected a characterizing Littlewood-Paley theorem of the same type as in the previous sections.

Besides the criteria in Sect. 5.2 and 6.1, it turns out that a useful tool for estimating operators is the study of their matrices in the basis $\{\psi_I\}$. In Meyer (1987) one can find a detailed analysis of Calderón-Zygmund operators.

Theorem 4.60. *Let T be a bounded linear operator in $L^2(\mathbb{R})$. The following conditions are equivalent.*

(i) *T is a Calderón-Zygmund operator and $T1 = T^*1 = 0$.*

(ii) $\langle T\psi_I, \psi_J\rangle \le c|I|^{1/2+\varepsilon}|J|^{1/2+\varepsilon}(|I|+|J|)^{-\varepsilon}(|I|+|J|+\operatorname{dist}(I,J))^{-1-\varepsilon}$ *for some $\varepsilon > 0$.*

The operators in Theorem 4.60 form an algebra, that is, the product of two such operators is an operator of the same type. As in Sect. 5 and Sect. 6, one can with the aid of wavelets construct *paraproducts*. Let $\varphi \in \mathcal{D}(\mathbb{R})$ be given with $\operatorname{supp}\varphi$ contained in $[0,1]$ and $\int_0^1 \varphi(x)dx = 1$. Set

$$\varphi_I(x) = 2^j\varphi(2^jx - k), \quad I = I_{jk} \in \mathcal{G}.$$

For $b \in \mathrm{BMO}$ consider the operator

$$L_bf(x) = \sum_{I\in\mathcal{G}} \gamma(I)\langle f, \varphi_I\rangle\psi_I(x),$$

where $\gamma(I) = \langle b, \psi_I\rangle$ are the coefficients of b.

Theorem 4.61. *If $b \in \mathrm{BMO}$ then the operator L_b is Calderón-Zygmund and*

$$L_b1 = b, \quad L_b^*1 = 0.$$

Chapter 5
Applications to the Theory of Singular Integrals

As an immediate application of the Littlewood-Paley theory we get a proof of the $T1$-Theorem and the Tb-Theorem, which give a general criterion for the boundedness of singular integrals in L^2. In Sect. 1 we set forth the necessary definitions. In particular, we discuss the notion of weak boundedness of an operator, formulate the $T1$-Theorem and the Tb-Theorem, and, finally, indicate the plan for their proofs. As an example we explain what these criteria give in the case of the L^2-boundedness problem for the Cauchy integral on Lipschitz curves.

The Cauchy integral on Lipschitz curves will be treated specially in Sect. 2. We begin with a general theorem of the boundedness of Calderón commutators, which turns out to be equivalent to the L^2-boundedness of the Cauchy integral. Then we analyze various proofs of this boundedness: Calderón's original proof in 1977, the Coifman-McIntosh-Meyer proof, and the perturbation proof by David and Murai. We remark also that sharp estimates for the norm of the Cauchy integral in terms of the Lipschitz constant can be obtained especially by the last path.

Section 3 is devoted to David's theorem. This gives a complete description of the curves on which the Cauchy integral is bounded in L^2: these are the Carleson curves. In the final Sect. 3.6 we discuss a counterexample, likewise due to David: the Cauchy integral on a Cantor set.

Finally, the concluding Sect. 4 is almost not connected at all with the rest of the Chapter. Here we have collected results which by various reasons were not treated in Chapters 2 and 3, but which may be useful in the applications or in the teaching of the theory of singular integrals.

The divison of the material of this part has to a high extent been formal. It is easy to see that the ideas and theorems of Chapters 4 and 5 intermingle and have evolved simultaneously. Stein (1970a) writes: " ... we have purposely not chosen the shortest and most direct way; we hope, however, that the longer route we shall follow will be more instructive. In this way the reader will have a better opportunity to examine all the working parts of the complex mechanism detailed below".

§1. Weak Boundedness

1.1. Singular Integral Operators. Recall (Sect. 1.1 of Chapter 3) that a Calderón-Zygmund kernel in $\mathbb{R}^n$ is a kernel $K(x,y)$, $x, y \in \mathbb{R}^n$, $x \neq y$, such that

$$|K(x,y)| \leq \frac{c}{|x-y|^n}, \tag{5.1}$$

$$|K(x,y) - K(x',y)| \leq c\frac{|x-x'|^\alpha}{|x-y|^{n+\alpha}}, \quad |x-x'| < \frac{1}{2}|x-y|, \tag{5.2}$$

$$|K(x,y) - K(x,y')| \leq c\frac{|y-y'|^\alpha}{|x-y|^{n+\alpha}}, \quad |y-y'| < \frac{1}{2}|x-y|, \tag{5.3}$$

Here $0 < \alpha \leq 1$. In Sect. 1.1 of Chapter 3 we have defined the Calderón-Zygmund operator with kernel K as the bounded linear operator in $L^2(\mathbb{R}^n)$ such that

$$Tf(x) = \int_{\mathbb{R}^n} K(x,y)f(y)dy$$

for all $f \in \mathcal{D}(\mathbb{R}^n)$ and $x \notin \operatorname{supp} f$. This relation can easly be rewritten in "weak" terms:

$$\langle Tf, g\rangle = \iint_{\mathbb{R}^n \times \mathbb{R}^n} K(x,y)f(y)g(x)dxdy \tag{5.4}$$

for arbitrary $f, g \in \mathcal{D}(\mathbb{R}^n)$, $\operatorname{supp} f \cap \operatorname{supp} g = \emptyset$.

But in (5.4) it is not necessary that Tf should be defined as an L^2-function! It suffices to take Tf as a distribution, i. e. a continuous linear functional on $\mathcal{D}(\mathbb{R}^n)$. Then the inner product $\langle Tf, g\rangle$ makes sense and we can ask whether (5.4) is true or not. We are led to the following definition.

Definition. A continuous linear operator $T : \mathcal{D}(\mathbb{R}^n) \to \mathcal{D}'(\mathbb{R}^n)$ is said to be a *singular integral operator* if there exists a Calderón-Zygmund kernel K with which it is connected via formula (5.4). The set of such operators will be written SIO_α.

If a singular integral operator is bounded in L^2-norm, then it is a Calderón-Zygmund operator. But our definition comprises also all differential operators with L^∞-coefficients; then the kernel equals 0.

Let us consider the family of seminorms

$$\|f\|_{B,m} = \sum_{|\beta|\le m} r^{|\beta|} \max_B |\partial^\beta f|, \quad \operatorname{supp} f \subset B,$$

where $B = B(x,r)$ is an arbitrary ball in $\mathbb{R}^n$ and $m \ge 0$ is an integer.

Definition. A singular integral operator T is *weakly bounded* if for some m holds

$$|\langle Tf, g\rangle| \le c|B|\, \|f\|_{B,m}\|g\|_{B,m} \tag{5.5}$$

for all $f, g \in \mathcal{D}$, $\operatorname{supp} f \cup \operatorname{supp} g \subset B$. It is readily seen that Calderón-Zygmund operators meet this condition: if T is bounded in L^2 then

$$|\langle Tf, g\rangle| \le c\|f\|_{L^2}\|g\|_{L^2} \le c|B|\, \|f\|_{B,0}\|g\|_{B,0}.$$

The preceding definition of weak boundedness is equivalent to the following more abstract one: for arbitrary bounded sets[2]

$$\mathcal{B}_1 \subset \mathcal{D} \quad \text{and} \quad \mathcal{B}_2 \subset \mathcal{D}$$

holds

$$|\langle Tf^{x,t}, g^{x,t}\rangle| \le c(\mathcal{B}_1, \mathcal{B}_2)\, t^n,$$

where $f \in \mathcal{B}_1$, $g \in \mathcal{B}_2$ and $f^{x,t}(y) \stackrel{\text{def}}{=} f(\frac{x-y}{t})$.

Condition (5.5) is much easier to check than boundedness in L^2. Let us give two examples.

Example 1. Let K be an antisymmetric Calderón-Zygmund kernel:

$$K(x,y) = -K(y,x),$$

[2] Recall (Rudin (1973)) that the boundedness of $\mathcal{B} \subset \mathcal{D}$ means that all functions in $\mathcal{B}$ have their support in a fixed ball B and that they are uniformly bounded in $C^\infty(B)$.

and let the operator T be defined in the principal value sense:

$$\langle Tf, g\rangle = \lim_{\varepsilon\to 0}\iint_{|x-y|>\varepsilon} K(x,y)f(y)g(x)dxdy.$$

Then

$$\langle Tf, g\rangle = \frac{1}{2}\iint K(x,y)[f(y)g(x) - f(x)g(y)]dxdy;$$

this integral is absolutely convergent for $f, g \in \mathcal{D}$ and satisfies (5.5) with $m = 1$. The operator T is weakly bounded.

Example 2. Assume that $T(e^{ix\xi}) \in \mathrm{BMO}(\mathbb{R}^n)$ for each $\xi \in \mathbb{R}^n$ and that $\sup_\xi \|T(e^{ix\xi})\|_{\mathrm{BMO}} < \infty$. Then T is weakly bounded. Indeed, if $f \in \mathcal{D}$ and $\operatorname{supp} f \subset B$, then $|Tf(x)| \le c\|f\|_{B,0}$ for $x \notin 2B$ by the estimates for the kernel. On the other hand

$$\|f\|_{\mathrm{BMO}} \le \int \|T(e^{ix\xi})\|_{\mathrm{BMO}}|\hat{f}(\xi)|d\xi \le c\|f\|_{B,n+1}$$

so that

$$\int_B |Tf| \le c|B|\,\|f\|_{B,n+1},$$

which implies (5.5).

The singular integral operator is initially defined in $\mathcal{D}$. However one can reasonably define Tf also for $f \in C^\infty \cap L^\infty$ with noncompact support. Let $g \in \mathcal{D}$ with $\int_{\mathbb{R}^n} g dx = 0$. Define $\langle Tf, g\rangle$ by the formula

$$\langle Tf, g\rangle = \langle T(\varphi f), g\rangle + \iint K(x,y)[1-\varphi(y)]f(y)g(x)dxdy,$$

where $\varphi \in \mathcal{D}$ and $\varphi = 1$ near $\operatorname{supp} g$. It is easy to see that this expression does not depend on the choice of φ (cf. the renormalization in Sect. 4.3 of Chapter 2 and Sect. 5.6 of Chapter 3) and is well defined. Tf is a linear functional on the space $\{g \in \mathcal{D} : \int g = 0\}$, that is, an element of $\mathcal{D}'/\mathbb{C}$.

We defined the class SIO_α and weak boundedness starting with the space $\mathcal{D}(\mathbb{R}^n)$. But it is not always convenient to work with infinitely differentiable functions. For example, on spaces of homogeneous type and on Lipschitz submanifolds of $\mathbb{R}^n$ one cannot always define infinite differentiability, but just Hölder classes of low order. It turns out that the entire theory connected with weak boundedness can be built on the basis of spaces of low smoothness.

For example, let $0 < s < 1$ and consider in $\mathbb{R}^n$ the *Hölder class* $\Lambda^s(\mathbb{R}^n)$ consisting of functions f of finite norm

$$\|f\|_\Lambda = \sup_{x\ne y}\frac{|f(x)-f(y)|}{|x-y|^s}.$$

The subspace of functions in Λ^s consisting of all functions with compact supports will be denoted Λ_0^s. Let $[\Lambda_0^s]'$ be the dual space of distributions. Now we can define a *singular integral operator* T as a linear operator from Λ_0^s into $[\Lambda_0^s]'$ such that (5.4) holds, with $f, g \in \Lambda_0^s$ with nonintersecting supports. The definition of Tf extends in a natural way to functions $f \in \Lambda^s \cap L^\infty$ with arbitrary support. This is a linear function in $[\Lambda_0^s]'/\mathbb{C}$.

Let us now formulate weak boundedness in the language of Λ^s. The semi-norm $\|f\|_{B,m}$ is invariant under dilation; it remains unchanged if we replace $f(x)$ by $f(\lambda x)$, $\lambda > 0$, and, at the same time, B by $\lambda^{-1}B$. Therefore the correct analogue of $\|f\|_{B,m}$ is not $\|f\|_{\Lambda^s}$ but rather $r^s\|f\|_{\Lambda^s}$.

The singular integral operator $T : \Lambda_0^s \to [\Lambda_0^s]'$ is said to be *weakly bounded* if

$$|\langle Tf, g\rangle| \le c|B|r^{2s}\|f\|_{\Lambda^s}\|g\|_{\Lambda^s} \tag{5.6}$$

when $f, g \in \Lambda_0^s$ with $\operatorname{supp} f \cap \operatorname{supp} g = \emptyset$.

It is clear that in $\mathbb{R}^n$ it is sufficient to verify (5.6) for functions in $\mathcal{D}$ and that (5.6) follows from (5.5). Also the converse is true, so that our two definitions of weak boundedness are equivalent.

Lemma 5.1. *If the operator $T : \mathcal{D} \to \mathcal{D}'$, $T \in \mathrm{SIO}_\alpha$, is weakly bounded in the first sense then (5.6) holds true for $f, g \in \mathcal{D}$.*

The proof of this lemma (David, Journé, and Semmes (1985), (1986)) is a good example of the technique of integral representations in Sect. 5 of Chapter 4. Let $\psi \in \mathcal{D}(\mathbb{R}^n)$ be radial symmetric satisfying condition (4.21) and $\operatorname{supp}\psi \subset B(0,1)$. In the notation of Sect. 5 of Chapter 4

$$\langle Tf, g\rangle = c_0^{-2} \iint_{\mathbf{R}_+^{n+1}\times\mathbf{R}_+^{n+1}} L(x,t;y,s)Q_tf(x)Q_tg(y)\frac{dx\,dt}{t}\frac{dy\,ds}{s},$$

where

$$L(x,t;y,s) = \langle T\psi_t(\cdot - x), \psi_s(\cdot - y)\rangle \quad \text{(formula (4.28))}.$$

Let us estimate L. If $|x-y| \ge 2(s+t)$, then in view of the estimate (5.1) for the kernel of T

$$|L(x,t;y,s)| \le \frac{c}{|x-y|^n}.$$

In order to estimate L for $|x-y| < 2(s+t)$ let us consider a function $\eta \in \mathcal{D}$ such that $0 \le \eta \le 1$, $\eta = 1$ in $B(x,2t)$, $\eta = 0$ off $B(x,3t)$ and $|\nabla\eta| \le ct^{-1}$. For $s \le t$ we have

$$L = \langle T\psi_t(\cdot - x), \eta\psi_s(\cdot - y)\rangle + \langle T\psi_t(\cdot - x), [1-\eta]\psi_s(\cdot - y)\rangle.$$

In the second term the supports of $\psi_t(\cdot - x)$ and $[1-\eta]\psi_s(\cdot - y)$ do not intersect and, in view of (5.4), it does not exceed $cs^{-n}\log 5s/2t$. In the first term we use weak boundedness:

$$|\langle T\psi_t, \eta\psi_s\rangle| \le c\cdot s^{-n}.$$

Finally, for $|x-y| < 2(s+t)$

$$|L(x,t;y,s)| \le \frac{c}{(s+t)^n}\left(1+\left|\log\frac{s}{t}\right|\right).$$

On the other hand, it is easy to estimate $Q_t f$ by $\|f\|_{\Lambda^s}$: if $\operatorname{supp} f \subset B(x_0,r)$ holds $Q_t f(x) = 0$ for $|x-x_0| > t+r$, while for $|x-x_0| < 2r$ and $t < r$ holds $|Q_t f(x)| \le ct^s\|f\|_{\Lambda^s}$. Thus, at any rate,

$$|Q_t f(x)| \le ct^{-n}|B|\,\|f\|_\infty \le ct^{-n}|B|r^s\|f\|_{\Lambda^s}.$$

Collecting all estimates, we obtain (5.6).

1.2. The $T1$-Theorem. In this section it will be convenient not to view elements of $\mathrm{BMO}(\mathbb{R}^n)$ as functions, but rather as equivalence classes *modulo* constants (as $\|1\|_{\mathrm{BMO}} = 0$). Then $\mathrm{BMO} \subset \mathcal{D}'/\mathbb{C}$.

The operator $T : \mathcal{D} \to \mathcal{D}'$ admits the conjugate operator $T^* : \mathcal{D} \to \mathcal{D}'$ with respect to the standard duality

$$\langle f, g\rangle = \int_{\mathbb{R}^n} f(x)g(x)dx.$$

It is defined by the identity $\langle T^* f, g\rangle = \langle f, Tg\rangle$ and if $T \in \mathrm{SIO}_\alpha$ then $T^* \in \mathrm{SIO}_\alpha$ too. The kernel of T^* is $K(y,x)$.

Theorem 5.2. *Let T be a weakly bounded operator in $T \in \mathrm{SIO}_\alpha$. It is bounded in L^2 (that is, a Calderón-Zygmund operator) if and only if*

$$T1 \in \mathrm{BMO}, \quad T^*1 \in \mathrm{BMO}.$$

Theorem 5.2 is called the $T1$-*Theorem*. It was proved by David and Journé (1984). Other proofs of Theorem $T1$, as well as generalizations and applications to the Cauchy integral, can be found in David (1986), David, Journé, and Semmes (1985), (1986), Meyer (1987).

If T is a Calderón-Zygmund operator, then it acts from L^∞ into BMO (Theorem 3.17), so that the condition in Theorem 5.2 is necessary.

The proof of the sufficiency starts with the construction of the paraproduct in Sect. 5.4 of Chapter 4. For any two functions b_1 and b_2 in BMO one constructs there a Calderón-Zygmund operator $T_0 \in \mathrm{SIO}_1$ such that $T_0 1 = b_1$, $T_0^* 1 = b_2$. Then taking $b_1 = T1$, $b_2 = T^*1$ and setting $T_1 = T - T_0$ we get $T_1 1 = T_1^* 1 = 0$. Therefore it is sufficient to consider the case $T1 = T^*1 = 0$ in Theorem 5.1.

Let once more $\psi \in \mathcal{D}(\mathbb{R}^n)$ be radially symmetric satisfying condition (4.21) and with $\operatorname{supp}\psi \subset B(0,1)$. Utilizing Lemma 4.36 we see that T will be bounded in L^2 if we can show that

$$\sup_{x,t}\int |L(x,t;y,s)|\frac{dyds}{s} + \sup_{y,s}\int |L(x,t;y,s)|\frac{dxdt}{t} < \infty, \tag{5.7}$$

where L is defined by formula (4.28):

$$L(x,t;y,s) = \langle T\psi_t^x, \psi_s^y\rangle, \quad \psi_t^x(z) = \psi_t(z-x).$$

The estimate (5.7) is easily proved: we have to estimate L carefully in various domains of $\mathbb{R}_+^{n+1} \times \mathbb{R}_+^{n+1}$, taking account the ratios between $|x-y|$, t and s. First we consider the domain

$$\Omega = \left\{(x,t;y,s) : |x-y| \le \frac{t}{10},\ s \le \frac{t}{10}\right\}.$$

Let again (cf. Sect. 1.1) η be a cut-off function for the ball $B(y,2s)$, that is, $\eta \in \mathcal{D}$, $\operatorname{supp}\eta \subset B(y,3s)$, $\eta = 1$ in $B(y,2s)$ and $|\partial^\beta \eta| \le c_\beta s^{-|\beta|}$. Then

$$L = \langle T\{\psi_t^x(\cdot) - \psi_t^x(y)\}, \psi_s^y\rangle,$$

because $T1 = 0$, $\int_{\mathbf{R}^n} \psi_s = 0$. Furthermore,

$$T\{\psi_t^x - \psi_t^x(y)\} = T[\eta\{\psi_t^x - \psi_t^x(y)\}] + T[(1-\eta)\{\psi_t^x - \psi_t^x(y)\}].$$

In the second term the support of the integrand does not meet the support of ψ_y^s and its contribution to L equals

$$\int_{\mathbf{R}_+^{n+1}\times\mathbf{R}_+^{n+1}} \psi_s(u-y)[K(u,v) - K(y,v)][1-\eta(v)][\psi_t(v-x) - \psi_t(y-x)]dudv.$$

Here we can leave out the term $K(y,v)$, because $\int_{\mathbf{R}^n} \psi_s(u-y)du = 0$ so that the result does not change. Now the estimate (5.2) for the kernel shows that this expression does not exceed $c \cdot s^\alpha \cdot t^{-n-\alpha}$.

The contribution to L of the first term

$$\langle T[\eta\{\psi_t^x - \psi_t^x(y)\}], \psi_s^y\rangle$$

is in view of the weak boundedness not greater than $c \cdot s \cdot t^{-n-1}$. Altogether, we have in the domain Ω

$$|L(x,t;y,s)| \le c\frac{s^\alpha}{t^{n+\alpha}},$$

so that, for example,

$$\int_\Omega |L|\,\frac{dxdt}{t} \le cs^\alpha \int_{10s}^\infty \frac{dt}{t^{n+1+\alpha}} \int_{|y-x|<10t} dx \le cs^\alpha \int_{10s}^\infty \frac{dt}{t^{1+\alpha}} \le B < \infty.$$

In exactly the same way one estimates L in other domains. For instance, if $t < s/10$ one has to use instead of $T1 = 0$ the condition $T^*1 = 0$.

Remarks. 1) Instead of Lemma 4.36 one can use other variants of Littlewood-Paley theory, for example, Coifman's construction in Sect. 6 of Chapter 4. Then, in view of Lemma 4.43 it suffices to check that in the assumptions of the $T1$-Theorem ($T1 = T^*1 = 0$) the matrix

$$a_{jk} = \|D_j T D_k\|$$

defines a bounded linear operator in $l^2(\mathbb{Z})$. Again, the kernel $E_{jk}(x, y)$ can be written down explicitly (this is the analogue of L) and one can, in a similar way as above, verify that

$$\int_{\mathbb{R}^n} |E_{jk}(x,y)|dx + \int_{\mathbb{R}^n} |E_{jk}(x,y)|dy \le c \cdot 2^{-\alpha|j-k|}.$$

From this follows immediately that $\|D_j T D_k\| \le c \cdot 2^{-\alpha|j-k|}$, so that T is bounded in $L^2(\mathbb{R}^n)$. In Sect. 6.3 of Chapter 4 we have given the construction of the corresponding paraproducts, which reduce the $T1$-Theorem to the case $T1 = T^*1 = 0$.

This proof depends neither on $\mathcal{D}(\mathbb{R}^n)$ nor Fourier analysis. Therefore it carries over to the case of spaces of homogeneous type.

2) The original proof by David and Journé (1984) was based on the Cotlar-Stein Lemma 3.6. But in the first step of the proof they were forced to rely on Littlewood-Paley theory for the construction of the paraproducts.

1.3. The Tb-Theorem. We have seen that of the two conditions in the $T1$-Theorem one is easy to check, namely the weak boundedness of the operator in question. However, the condition $T1 \in \mathrm{BMO}$ usually requires ań explicit computation of $T1(x)$. Therefore the $T1$-Theorem is not sufficiently adaptable for playing the rôle of an effecient L^2-boundedness criterion.

Let b be in $L^\infty(\mathbb{R}^n)$. Is it possible to use Tb instead of $T1$? The first result of this kind is due to McIntosh and Meyer (1985). Thus let b be in $L^\infty(\mathbb{R}^n)$ and assume that $Tb = T^*b = 0$. McIntosh and Meyer showed that if $\mathrm{Re}\, b \ge \varepsilon > 0$ in $\mathbb{R}^n$ then the weak boundedness of the operator

$$M_b T M_b : f \mapsto b \cdot T(bf)$$

entails the boundedness of T in $L^2(\mathbb{R}^n)$.

Recently David, Journé, and Semmes (1985), (1986) have obtained an exact generalization in this direction of the $T1$-Theorem, which is called the Tb-Theorem.

As we now have to apply the singular integral operator T to a nonsmooth function such as b or bf, with $f \in \mathcal{D}$, it is necessary to change somewhat the definition of singular integral operator.

Let $b_1, b_2 \in L^\infty(\mathbb{R}^n)$. By a *singular integral operator* we mean a continuous linear operator[3]

$$T : b_1\mathcal{D}(\mathbb{R}^n) \to [b_2\mathcal{D}(\mathbb{R}^n)]'$$

[3] The topology in $b\mathcal{D}$ is derived from the one in $\mathcal{D}$.

such that with a suitable Calderón-Zygmund kernel K holds

$$\langle Tb_1 f, b_2 g\rangle = \iint_{\mathbf{R}^n \times \mathbf{R}^n} g(x) b_2(x) K(x,y) b_1(y) f(y) dx dy, \tag{5.8}$$

where, as usual, $f, g \in \mathcal{D}$ and $\operatorname{supp} f \cap \operatorname{supp} g = \emptyset$.

Here $b\mathcal{D} \overset{\text{def}}{=} \{bf : f \in \mathcal{D}\}$. Let M_b be the operator of pointwise multiplication with the function b. Then

$$\langle Tb_1 f, b_2 g\rangle = \langle M_{b_2} T M_{b_1} f, g\rangle,$$

but the operator $M_{b_2} T M_{b_1} f$ has the kernel $b_2(x)K(x,y)b_1(y)$, which does not satisfy the estimates (5.2) and (5.3). Nevertheless, estimate (5.1) remains in force, and it is possible to speak of the weak boundedness of $M_{b_2} T M_{b_1} f$, regardless of whether it is question of $\mathcal{D}$ or Λ_0^s.

Finally, in analogy to Sect. 1.1, we define $T(b_1 f)$ for $f \in C^\infty \cap L^\infty$ or for $f \in \Lambda^s \cap L^\infty$ with arbitrary support. This a continuous linear functional on the subspace

$$\left\{ b_2 g; g \in \mathcal{D}, \int_{\mathbf{R}^n} b_2 g = 0 \right\},$$

that is, an element of $[b_2\mathcal{D}]'/\mathbb{C}$ (or of $[b_2\Lambda_0^s]'/\mathbb{C}$).

If T is bounded in L^2 then $M_{b_2} T M_{b_1}$ is, of course, weakly bounded (and indeed bounded in L^2!), while $Tb_1 \in \mathrm{BMO}$, $T^* b_2 \in \mathrm{BMO}$ for any interpretation of the action of T on b.

The following converse statement, which may be fulfilled or may not be fulfilled for a given pair of functions b_1, b_2, is called the *Tb-Theorem.*

The Tb-Theorem. *Let T be a singular integral operator in $\mathbb{R}^n$. If $M_{b_2} T M_{b_1}$ is weakly bounded, if $Tb_1 \in \mathrm{BMO}$ and $T^* b_2 \in \mathrm{BMO}$, then T is bounded in $L^2(\mathbb{R}^n)$.*

The $T1$-Theorem (Theorem 5.2) is obtained as the special case $b_1 = b_2 = 1$.

David, Journé, and Semmes (1985), (1986) have recently described those functions for which the Tb-Theorem holds.

If $b_1 = b_2 = b$, then clearly the operation of multiplication with b^{-1} satisfies all the assumptions of the Tb-Theorem. However, it is bounded in L^2 only if $b^{-1} \in L^\infty$. Therefore, we will assume in what follows that $b_1^{-1}, b_2^{-1} \in L^\infty(\mathbb{R}^n)$.

Definition. A function $b \in L^\infty(\mathbb{R}^n)$ is called *para-accretive* if any ball $B = B(x,r) \subset \mathbb{R}^n$ contains a smaller ball $B_1 = B(y,s) \subset B$ such that $s > r/N$ and

$$\left| \frac{1}{|B_1|} \int_{B_1} b(x) dx \right| \geq \varepsilon > 0. \tag{5.9}$$

Here N and ε depend only on b and not on the ball.

Theorem 5.3. (i) *Let b_1 and b_2 be para-accretive. Then the Tb-Theorem is fulfilled for the pair b_1, b_2.*

(ii) *If the Tb-Theorem is fulfilled for $b_1 = b_2 = b$ then b must be para-accretive.*

Let us give some sufficient conditions for para-accretiveness.

If b is *accretive*, that is, $\operatorname{Re} b \geq \varepsilon > 0$ then (5.9) is fulfilled, of course, and the McIntosh-Meyer theorem (1985) follows from Theorem 5.3.

Next, let $\{s_k(x,y)\}_{-\infty<k<\infty}$ be a quasi-identity in $\mathbb{R}^n$ (cf. Sect. 6.1 of Chapter 4). We say that a function b is *pseudo-accretive* if for any k

$$\left|\int_{\mathbb{R}^n} s_k(x,y)b(y)dy\right| \geq \varepsilon > 0. \tag{5.10}$$

It is clear that every pseudo-accretive function is para-accretive.

One can show that if $\varphi \in \mathrm{BMO}$ and $\|\varphi\|_{\mathrm{BMO}}$ is sufficiently small, then the function $b = e^{i\varphi}$ is pseudo-accretive. If $z = z(s)$ is the natural parametrization of a planar curve, with the arc commensurable with the chorde, then the function $z'(s)$ is pseudo-accretive. On the other hand, the function e^{ix} in $\mathbb{R}^1$ is not even para-accretive, as its average along any sufficiently large interval is arbitrarily small.

Besides, in $\mathbb{R}^1$ every para-accretive function is also pseudo-accretive (David, Journé, and Semmes (1985), (1986)). In $\mathbb{R}^2$ this is not so because of purely topological reasons. Consider, for example, the function $b(x,y) = \frac{x+iy}{|x+iy|}$ in the plane. This function is para-accretive, but if there were a quasi-identity enjoying property (5.10) then the function

$$u_R(e^{i\theta}) = \int_{\mathbb{R}^2} s_0[Re^{i\theta}, z]b(z)dz$$

would for R sufficiently large be as close (in the uniform norm) to the function $e^{i\theta}$ as we wish. In other words, the winding number of the complex vector field

$$x \mapsto \int_{\mathbb{R}^2} s_0(x,y)b(y)dy$$

over the circumference $\{x = Re^{i\theta}\}$ equals 1, but then it must vanish somewhere in the disk $\{|x| < R\}$.

The proof of Theorem 5.3 follows the scheme of proof of the $T1$-Theorem in Sect. 1.2 (Coifman's approach). First David, Journé, and Semmes (1985), (1986) prove the Tb-Theorem not for para-accretive functions but for pseudo-accretive functions. Let $\{S_k^*\}$ be the quasi-identity corresponding to b, $\{S_k\}$ and $\{D_k\}$ being the operators in Sect. 6.1 of Chapter 4. Set

$$P_k = S_k^*\{S_k b\}^{-1} S_k M_b,$$

so that the kernel of the operator P_k equals

$$p_k(x,y) = b(y)\int_{\mathbb{R}^n} \frac{s_k(z,x)s_k(z,y)}{S_k b(z)}dz.$$

Then $P_k 1 = 1$ and $P_K^* b = b$. Set

$$E_k = P_k - P_{k-1}.$$

It turns out that the operators $\{E_k\}$ in many respects have properties similar to those of the $\{D_k\}$. In particular,

$$\sum_{-\infty}^{\infty} \|E_k f\|_{L^2}^2 + \|E_k^* f\|_{L^2}^2 \le c\|f\|_{L^2}^2, \quad f \in L^2(\mathbb{R}^n), \tag{5.11}$$

After that one constructs the operator

$$T_n = \sum_{|j-k|\le n} E_j E_k$$

and, as in Sect. 6 of Chapter 4, it turns out that it is invertible in L^2 and in BMO for sufficiently large n. This construction is extended to b_1 and b_2 so that as a consequence one gets operators E_{j1}, E_{j2}, T_{n1} and T_{n2}. If $Tb_1 = T^* b_2 = 0$ then instead of the estimates for T analogous to Lemma 4.43 one proves the boundedness of the operator $T_{n2} T T_{n1}$, for which one has the estimate

$$\|E_{j2} T E_{k1}^*\| \le c \cdot 2^{-\nu|j-k|}$$

with a suitable $\nu > 0$. The last estimate is gotten from direct pointwise estimates of the kernel of the operator $E_{j2} T E_{k1}^*$.

The general case is reduced to the special case $Tb_1 = T^* b_2 = 0$, exactly as in Sect. 1.2, with the aid of paraproducts. In analogy with Sect. 6.3, if $\beta = Tb_1 \in \mathrm{BMO}$, $\gamma = T_{n2}^{-1}\beta \in \mathrm{BMO}$, we have

$$L_1 = \sum_k E_{k2}\{E_{k2}^n \gamma\} P_{k1} M_{b_1}^{-1}.$$

This is a Calderón-Zygmund operator (cf. Sect. 6.3) and $L_1 b_1 = T_{n2}\gamma = \beta$, $L_1^* b_2 = 0$ (as $P_{k1} 1 = 1$ and $E_{k2}^* b_2 = 0$). The operator L_2 is gotten in an analogous fashion and then it remains to prove, instead for T, the boundedness for $T - L_1 - L_2$.

For para-accretive functions one has to imitate the above proof, but instead of $\{s_k\}$ one constructs averaging kernels $v_k(x, y)$ which are smooth in y but not in x. Instead of smoothness in x this kernel enjoys the property of being constant in each dyadic cube in $\mathbb{R}^n$ of side length 2^{-k-N}, where N is sufficiently big.

To the expense of this one manages to rescue the estimate from below (5.10) with v_k instead of s_k, and to finish the proof. The execution this program in practise – in particular the derivation of (5.11) – is connected with great technical difficulties. The complete proof of the Tb-Theorem can be found in David, Journé, and Semmes (1985), (1986).

The Tb-Theorem and its proof extend easily to spaces of homogeneous type (in which case one has of course to depart from Λ_0^s and not $\mathcal{D}$) with an infinite nonatomic measure. There exists likewise a generalization to the vectorial case, when b_1, b_2 and K are matrix functions (Brackx, Delanghe, and Sommer (1982), David, Journé, and Semmes (1985), (1986)).

1.4. Application: The Cauchy Integral. Let us apply the $T1$- and the Tb-Theorem to the boundedness problem for the Cauchy integral on Lipschitz curves (cf. Sect. 2.4 of Chapter 4). In Sect. 2.4 of Chapter 3 we have encountered the singular integral operator

$$Tf(x) = \text{P.V.} \int_{-\infty}^{\infty} \frac{1}{x - y + i[\varphi(x) - \varphi(y)]} f(y)dy \tag{5.12}$$

with a Lipschitz function φ on $\mathbb{R}^1$. The boundedness of T in L^2 is equivalent to the boundedness in L^2 of the Cauchy integral of the curve with equation $y = \varphi(x)$. The kernel of T is anti-symmetric so that it is automatically weakly bounded (Sect. 1.1). But it is not possible to check that $T1 \in \text{BMO}$ for this operator so that the $T1$-Theorem does not give anything in the case of (5.12). On the other hand, we have a. e.

$$T(1 + i\varphi') = \pi i$$

(cf. Sect. 6.1 of Chapter 2) so that in $\mathcal{D}'/\mathbb{C}$ holds

$$T(1 + i\varphi') = T^*(1 + i\varphi') = 0.$$

The bounded function $b = 1 + i\varphi$ in $\mathbb{R}^1$ clearly is accretive. On the other hand, the kernel of $M_b T M_b$ is by the above anti-symmetric and satisfies the estimate (5.1). Consequently, it is weakly bounded, which shows that T is bounded in L^2. This reasoning was the main stimulus for the work of McIntosh and Meyer (1985), where for the first time a special case of the Tb-Theorem appeared.

But it is nevertheless possible to obtain information about the operator (5.12) from the $T1$-Theorem. Assume that the Lipschitz constant of φ satisfies $M = \|\varphi'\|_{L^\infty} < 1$. Then the operator (5.12) can be developed in a series of multiple commutators:

$$Tf = \sum_0^\infty (-i)^k T_k f = \sum_0^\infty \text{P.V.} \int_{-\infty}^{\infty} \left(\frac{\varphi(x) - \varphi(y)}{x - y} \right)^k \frac{f(y)}{x - y} dy.$$

Each of the commutators T_k has an anti-symmetric Calderón-Zygmund kernel with constant ckM^k. The operator T_0 coincides with the Hilbert transform and, consequently, is bounded. But directly by partial integration one can check that

$$T_{k+1}1 = T_k(\varphi').$$

By induction we find that all the commutators T_k are bounded in L^2 and that $\|T_k\| \leq c_1^{k+1} M^k$. Indeeed, if this estimate is true for T_k, then T_k must be a Calderón-Zygmund operator and therefore it follows that it acts from L^∞ into BMO with $\|T_k(\varphi')\|_{\mathrm{BMO}} \leq c_1^{k+1} M^k \|\varphi'\|_\infty \leq c_1^{k+1} M^k$. Hence the $T1$-Theorem is applicable to T_{k+1}: it is bounded in L^2 and from the proof of $T1$-Theorem it is clear that

$$\|T_{k+1}\| \leq c_1 (ckM^{k+1} + \|T_{k+1}1\|_{\mathrm{BMO}} + \|T_{k+1}^* 1\|_{\mathrm{BMO}}) \leq c_1^{k+2} M^{k+1}.$$

Inserting these estimates in the series for T we obtain

$$\|T\| \leq \sum_0^\infty c_1^{k+1} M^k < \infty,$$

provided $M < 1/c_1$. The $T1$-Theorem allows us to establish the boundedness of the Cauchy integral, but only for curves with sufficiently small Lipschitz constant and by series expansion. On the other hand the Tb-Theorem gives the boundedness of the Cauchy integral in one single step without any restriction on M.

§2. Calderón Commutators and the Cauchy Integral

2.1. The Commutator Theorem. Let φ be a real Lipschitz function on $\mathbb{R}^1$ with Lipschitz constant $M = \|\varphi'\|_{L^\infty}$. Let $h \in C^\infty(\mathbb{R}^1)$. By a *Calderón commutator* we mean the singular integral operator

$$T[h, \varphi]f = \text{P.V.} \int_{-\infty}^{\infty} h \left\{ \frac{\varphi(x) - \varphi(y)}{x - y} \right\} \frac{f(y)}{x - y} dy.$$

If $h(t) \equiv 1$ we obtain the Hilbert transform and if $h(t) = (1 + it)^{-1}$ the operator (5.12), i. e. the Cauchy integral for the curve $y = \varphi(x)$, while $h(t) = t^k$ gives the multiple commutators T_k of Sect. 1.4.

Theorem 5.4. *The singular integral operator $T[h, \varphi]$ is bounded in $L^2(\mathbb{R})$ and is a Calderón-Zygmund operator. For a suitable $m < \infty$ we have*

$$\|T\| \leq c \sum_{k=0}^{m} \sup_{[-M,M]} |h^{(k)}(t)|.$$

Let us mention three special cases of Theorem 5.4.

1) $h(t) = h_\lambda(t) = (1 + i\lambda t)^{-1}$, $\lambda \in \mathbb{R}$.

Then Theorem 5.4 says that

$$\|T[h_\lambda, \varphi]\| \leq c(1 + |\lambda M|)^m. \tag{5.13}$$

2) $h(t) = e^{i\xi t}$, $\xi \in \mathbb{R}$.
Theorem 5.4 gives

$$\|T[h,\varphi]\| \le c(1+|\xi M|)^m. \tag{5.14}$$

3) $h(t) = t^k$, so that $T[h,\varphi] = T_k$.
Theorem 5.4 gives

$$\|T_k\| \le ck^m M^k. \tag{5.15}$$

It turns out that to prove Theorem 5.4 itself it suffices to establish any of these special cases. Assume, for example, that the estimate (5.13) is known. Consider the function $h_\zeta = (\zeta - t)^{-1}$, where $\zeta = \xi + i\eta$, with $\eta \ne 0$, $|\xi| \le 2M$. It is easy to see that

$$T[h_\zeta,\varphi] = \frac{1}{i\eta} T[h_{1/\eta},\psi_\xi],$$

where $\psi_\xi(x) = \varphi(x) - \xi x$, so that $\|\psi_\xi'\|_{L^\infty} \le 3M$. Thus, in view of (5.13)

$$\|T[h_\zeta,\varphi]\| \le c\left(1 + \frac{M^m}{\eta^{m+1}}\right).$$

Let now $h \in C^\infty(\mathbb{R})$. Without changing the operator $T[h,\varphi]$ we may assume that $h(t) = 0$ for $|t| > 2M$ and that

$$\sum_{k=0}^{m+2} \|h^{(k)}\|_{L^\infty} \le c \sum_{k=0}^{m+2} \sup_{[-M,M]} |h^{(k)}(t)|.$$

Set (cf. Dyn'kin (1981))

$$F(x,y) = \sum_{k=0}^{m+1} h^{(k)}(x) \frac{(iy)^k}{k!}.$$

Then $F(x,0) = h(x)$ and

$$\frac{\partial F}{\partial \bar z} = \frac{1}{2}\left(\frac{\partial F}{\partial x} + i\frac{\partial F}{\partial y}\right) = \frac{1}{2} h^{(m+2)}(x) \frac{(iy)^{m+1}}{(m+1)!}.$$

Let Γ be a contour going around the segment $[-M,M]$ at the distance 1 and denote its interior by Γ^0. By the Cauchy-Green formula we have

$$h(x) = \frac{1}{2\pi i}\int_\Gamma \frac{F(\zeta)}{\zeta - x} d\zeta - \frac{1}{\pi}\iint_{\Gamma^0} \frac{\partial F}{\partial \bar\zeta}\frac{1}{\zeta - x} d\xi d\eta,$$

so that

$$T[h,\varphi] = \frac{1}{2\pi i}\int_\Gamma F(\zeta) T[h_\zeta,\varphi] d\zeta - \frac{1}{\pi}\iint_{\Gamma^0} \frac{\partial F}{\partial \bar\zeta} T[h_\zeta,\varphi] d\xi d\eta,$$

and this integral is absolutely convergent in operator norm in view of the estimates for $\|T[h_\zeta, \varphi]\|$. Hence $T[h, \varphi]$ is a bounded operator.

Exactly in the same way, if the estimate (5.14) is known, then instead of the Cauchy-Green formula we have to use the Fourier expansion

$$T[h, \varphi] = \frac{1}{\sqrt{2\pi}} \int_{-\infty}^{\infty} \hat{h}(\xi) T[e^{i\xi t}, \varphi] d\xi,$$

which is absolutely convergent if $h \in C^\infty$.

Finally, if (5.15) is known, then a not very difficult transformation (Coifman, McIntosh, and Meyer (1982)) allows one to pass to the estimate (5.13). Of course, as a result of all these reasonings we obtain some dependence on m and a deterioration of the constant c, in comparison to the original estimate (in particular, c depends on M).

But the estimate (5.13) is in fact already known to us! First, the operator $T[h_\lambda, \varphi]$ differs only by the circumstance that f has been replaced by $(1 + i\lambda\varphi')f$ from the Cauchy integral along the curve $\Gamma_\lambda = \{y = \lambda\varphi(x)\}$. If $\|T\|$ is the norm of T in $L^2(\mathbb{R})$ and $\|Q\|$ the norm of Q in $L^2(\Gamma_\lambda)$ (with respect to arc length), then

$$\|T\| \leq \|Q\| \quad \text{and} \quad \|Q\| \leq (1 + M)\|T\|. \tag{5.16}$$

In Sect. 2.4 of Chapter 4 we gave the Jones-Semmes proof of the boundedness of the Cauchy integral on Lipschitz curves and noted the estimate $\|Q\| \leq c(1 + M)^2$. Consequently, (5.13) holds with $m = 2$.

Second, in Sect. 1.4 we have just encountered another proof of the boundedness of the Cauchy integral with the aid of the Tb-Theorem. This proof likewise gives (5.12) with $m = 2$.

Thus, the commutator theorem is proved.

The rotation method (Sect. 4 in Chapter 3) allows us to deduce from the commutator theorem the following result.

Theorem 5.5. *Let φ be a Lipschitz function in $\mathbb{R}^n$ and let $h : \mathbb{R}^n \to \mathbb{R}^{n+1}$ be a mapping such that*

$$c_1|x - y| \leq |h(x) - h(y)| \leq c_2|x - y|.$$

Then the operator

$$Tf(x) = \text{P.V.} \int_{\mathbb{R}^n} \frac{\varphi(x) - \varphi(y)}{|h(x) - h(y)|^{n+1}} f(y) dy$$

is bounded in $L^2(\mathbb{R}^n)$.

Concerning applications of this theorem to double layer potentials and to the theory of boundary problems see Calderón (1980), Dahlberg and Kenig

(1985), Fabes, Jerison, and Kenig (1982), Fabes, Jodeit, and Lewis (1977), Fabes, Jodeit, and Rivière (1977), Verchota (1984).

2.2. The Question's History. In 1965 Calderón proved the L^2-boundedness of the commutator $T[t,\varphi]$. This result is set forth in Sect. 1.6 of Chapter 4 (Theorem 4.8). In Coifman and Meyer (1975) this is extended to the commutator $T[t^2,\varphi]$.

In a remarkable paper Calderón (1977) proved that the Cauchy integral is bounded in L^2 for $M < M_0$, where M_0 is an absolute constant.

In Sect. 1.4 we saw that this result follows from the $T1$-Theorem of David and Journé, which however appeared only in (1984). From this it follows that all the commutators $T[t^k,\varphi]$ are bounded, but not with a power-like estimate, just with an exponential one:

$$\|T[t^k,\varphi]\| \leq c^{k+1}M^k;$$

one likewise gets the boundedness of the operators $T[h,\varphi]$ for functions h which are holomorphic in the disk $\{|t| < M_0\}$.

In the same paper (Calderón (1977)) it is suggested that the restriction $M < M_0$ might not be essential, and Theorem 5.4 is formulated as a conjecture.

In Coifman, McIntosh, and Meyer (1982) the boundedness of the multiple commutators T_k is proved with the estimate (5.15) for $m = 4$. Almost immediately after that Theorem 5.4 was obtained in its full extent (Coifman, David, and Meyer (1983)). The proof in Coifman, McIntosh, and Meyer (1982) did not use Calderón's results but was rather difficult and troublesome. Coifman, Meyer, and Stein (1983) gave a simplified and more transparent variant.

In 1982 David proved a theorem concerning the boundedness of the Cauchy integral on Calderón curves, which we shall discuss in the following section. His method allows one to remove the restriction $M < M_0$ in a comparatively simple way, but David published his proof only in (1984). At the same time Murai (1983), (1984) managed to derive (5.14) from Calderón's theorem using about similar tools. In all these variants the proof is simpler than the one by Coifman, McIntosh, and Meyer.

The Tb-Theorem appeared first in McIntosh and Meyer (1985), and in its general form (Theorem 5.2) in David, Journé, and Semmes (1985), (1986). As a prototype for them served the $T1$-Theorem of David and Journé.

Finally, the Jones-Semmes proof appeared only in 1987.

2.3. Calderón's Proof. Calderón (1977) considered the Cauchy integral

$$T_\lambda f(x) = \text{P.V.}\int_{-\infty}^{\infty} \frac{1 + i\lambda\varphi'(y)}{x - y + i\lambda[\varphi(x) - \varphi(y)]} f(y)dy,$$

taken along the curve $\Gamma_\lambda = \{x + i\lambda\varphi(x), -\infty < x < \infty\}$, where $\lambda > 0$ is a parameter. For $\lambda = 0$ this operator coincides with the Hilbert transform and is bounded in $L^2(\mathbb{R})$. Let $\varphi \in C^\infty(\mathbb{R})$, $f \in \mathcal{D}(\mathbb{R})$. Calderón managed to estimate the derivative $dT_\lambda/d\lambda$.

Theorem 5.6.

$$\|dT_\lambda/d\lambda\| \le c\,(M+1)^3(1+\|T_\lambda\|^2).$$

From this it follows that the operator T_λ is bounded in L^2 for λ sufficiently small, while the operator T_1 is bounded for $M < M_0$, where M_1 is an absolute constant. Of course, the admissible λ's are those for which the ordinary differential equation

$$dy/d\lambda = c\,(M+1)^3(1+y^2)$$

has a finite solution. The derivative $dT_\lambda/d\lambda$ admits the following explicit expression:

$$(dT_\lambda/d\lambda)f(x) = \text{P.V.}\int_{-\infty}^{\infty} i\left\{\frac{\varphi(x)-\varphi(y)}{[x-y+i\lambda(\varphi(x)-\varphi(y))]^2} + \right.$$
$$\left. + \frac{1}{x-y+i\lambda(\varphi(x)-\varphi(y))}\cdot\frac{\varphi'(y)}{1+i\lambda\varphi'(y)}\right\}(1+i\lambda\varphi'(y))f(y)dy.$$

Hence,

$$\|dT_\lambda/d\lambda\| \le \|B_\lambda\| + M\|T_\lambda\|,$$

where

$$B_\lambda f(x) = \text{P.V.}\int_{-\infty}^{\infty}\frac{\varphi(x)-\varphi(y)}{[x-y+i\lambda(\varphi(x)-\varphi(y))]^2}(1+i\lambda\varphi'(y))f(y)dy.$$

Let us make a regularization of this singular integral, setting $z_\lambda(x) = x + i\lambda\varphi(x)$ and

$$D_\lambda f(x) = \lim_{\delta\to+0}\int_{-\infty}^{\infty}\frac{\varphi(x)-\varphi(y)}{[z_\lambda(x)-z_\lambda(y)+i\delta]^2}f(y)dz_\lambda(y).$$

Then it is readily seen that for $0 \le \lambda \le 1$

$$|D_\lambda f(x) - B_\lambda f(x)| \le cM\cdot Mf(x),$$

Mf being the Hardy-Littlewood maximal operator. Therefore

$$\|B_\lambda\| \le \|D_\lambda\| + cM.$$

Next, as in Sect. 1.6 of Chapter 4, let us represent $\varphi(x)-\varphi(y)$ in the form

$$\varphi(x)-\varphi(y) = \int_{-\infty}^{\infty}\varphi'(s)[e(x-s)-e(y-s)]ds, \quad e(t) = \chi_{(0,+\infty)}(t).$$

The curve Γ_λ divides the plane into two domains with Lipschitz boundary (an upper one G_+ and a lower one G_-). The Cauchy integral along Γ_λ (Sect. 6.1 of Chapter 2) allows us to represent f in the form

$$f(y) = f_+[z_\lambda(y)] - f_-[z_\lambda(y)],$$

where f_+ is analytic in G_+ and f_- analytic in G_-, and so by the definition of T_λ as a Cauchy integral (cf. (5.16)) we get

$$\|f_\pm\|_{E^2} \le c(M+1)(1+\|T_\lambda\|)\|f\|_{L^2}.$$

Finally, if g is another function in $\mathcal{D}(\mathbb{R})$, for which with have an analogous expansion in g_+ and g_-, then

$$\begin{aligned} L(f,g) &\stackrel{\text{def}}{=} \int_{-\infty}^{\infty} g(x) D_\lambda f(x) dz_\lambda(x) \\ &= \int_{-\infty}^{\infty} \varphi'(s)ds \left\{ f_+(z_\lambda(s))g_-(z_\lambda(s)) + \int_{\Gamma_{\lambda s}} [f'_+(\zeta)g_+(\zeta) + f'_-(\zeta)g_-(\zeta)]d\zeta \right\}, \end{aligned}$$

where $\Gamma_{\lambda s} = \{x + i\lambda\varphi(x), s < x < \infty\}$.

But by Calderón's theorem for Lipschitz curves (Theorem 4.13 (iii)) the expression within brackets is summable and

$$|L(f,g)| \le c(M+1)^3(1+\|T_\lambda\|^2)\|f\|_{L^2}\|g\|_{L^2}.$$

Hence

$$\|D_\lambda\| \le c(M+1)^3(1+\|T_\lambda\|^2),$$

establishing Theorem 5.6.

2.4. The Proof of Coifman-McIntosh-Meyer. The proof of the L^2-boundedness of the Cauchy integral on Lipschitz curves by the method of Coifman, McIntosh and Meyer is very complicated even in the simplified version given in Coifman, Meyer, and Stein (1983). We give here only the main steps.

The goal of this proof is the estimate (5.15) for the multiple commutators $T_k = T[t^k, \varphi]$.

Let $M_{\varphi'}$ stand for the multiplication operator with φ', and let $D = -i\frac{d}{dx}$ be differentiation and

$$P_t = (1+t^2D^2)^{-1}, \quad Q_t = tDP_t,$$

so that in $L^2(\mathbb{R})$

$$(1+itD)^{-1} = P_t - iQ_t. \tag{5.17}$$

Lemma 5.7.

$$T_k = \text{P.V.} \int_{-\infty}^{\infty} (1+itD)^{-1}[(1+itD)^{-1}M_{\varphi'}]^k \frac{dt}{t}. \tag{5.18}$$

This formula is known as the *integral representation of McIntosh* and is proved by a direct computation of the kernel of the operator in the right hand side of (5.18). It remains to be taken into account that after taking the Fourier transform the operator $(1+itD)^{-1}$ becomes multiplication by $(1+it\xi)^{-1}$ and its kernel equals $\frac{1}{|t|}k(\frac{x-y}{t})$, where $k(u)=e^{-u}$ for $u \geq 0$ and $k(t)=0$ for $u<0$.

Insert now (5.17) into (5.18) and remove the parantheses. The operator T_k comes then as a sum of terms of the following types:

1) The term with $(P_tM_{\varphi'})^kP_t$ drops out, because it is even in t.

2) $k+1$ terms contain the factor Q_t exactly once:

$$\text{P.V.}\int_{-\infty}^{\infty}(P_tM_{\varphi'})^pQ_t(P_tM_{\varphi'})^q\frac{dt}{t},\quad p+q=k. \tag{5.19}$$

3) Terms containing Q_t two times or more, combining into $k(k+1)$ blocks of the form

$$\text{P.V.}\int_{-\infty}^{\infty}(P_tM_{\varphi'})^pQ_t[M_{\varphi'}(1+itD)^{-1}\dots(1+itD)^{-1}M_{\varphi'}]Q_t(M_{\varphi'}P_t)^q\frac{dt}{t}. \tag{5.20}$$

One sees that the norm of any of the operators (5.19), (5.20) does not exceed $c(p+1)(q+1)\|\varphi'\|^k_{L^\infty}$, whence the estimate (5.15) with $m=4$.

The estimate of the operators (5.19) yields, with an elegant device, to an estimate of the operators (5.20). In this connection an important rôle is played by the operator identities

$$t\frac{\partial}{\partial t}P_t=-2Q_t^2,\quad t\frac{\partial}{\partial t}(2P_tQ_t-Q_t)=Q_t-8Q_t^3.$$

But the main part of the proof is the estimate for the operator (5.20).

Lemma 5.8. *Let $\{L_t\}$ be a family of operators in L^2 with $\|L_t\|\leq 1$. If the operator $\mathcal{L}$ is given by*

$$\mathcal{L}=\lim_{\varepsilon\to 0}\int_{\varepsilon}^{1/\varepsilon}(P_tM_{\varphi'})^pQ_tL_tQ_t(M_{\varphi'}P_t)^q\frac{dt}{t},$$

then

$$\|\mathcal{L}\|\leq c(p+1)(q+1)\|\varphi'\|^{p+1}_{L^\infty}.$$

Clearly Lemma 5.8 completes the proof. Let us now turn to the Littlewood-Paley theory in Sect. 5 of Chapter 4. If $H=L^2(\frac{dxdt}{t})$ in the upper halfplane $\mathbb{R}^2_+$ then, apparently,

$$\begin{aligned}|\langle\mathcal{L}f,g\rangle| &= \left|\int_0^\infty\langle L_tQ_t(M_{\varphi'}P_t)^qf,Q_t(M_{\varphi'}P_t)^pg\rangle\frac{dt}{t}\right|\\ &\leq \|Q_t(M_{\varphi'}P_t)^qf\|_H\,\|Q_t(M_{\varphi'}P_t)^pg\|_H.\end{aligned}$$

Lemma 5.9.

$$\|Q_t(M_{\varphi'}P_t)^{q+1}f\|_H \leq \|Q_t(M_{\varphi'}P_t)^q f\|_H + C\|f\|_{L^2}.$$

Clearly Lemma 5.8 will follow from this by induction, because $\|Q_t f\|_H \asymp \|f\|_{L^2}$ by Plancherel's formula. We have to take account of Theorem 4.38 of Sect. 5.3 of Chapter 4. Lemma 5.9 is readily derived from the following statement.

Lemma 5.10. *Let* $b \in L^\infty(\mathbb{R})$ *and let* $v(x,t) = v_t(x)$ *be a function in* $\mathbb{R}^2_+$ *such that* $\sup_t |v(x,t)| \in L^2(\mathbb{R})$. *Set*

$$\Delta(x,t) = P_t\{bQ_tv_t\}(x) - Q_t\{bP_tv_t\}(x).$$

Then

$$\|\Delta\|_H \leq c\|b\|_{L^\infty}\left\|\sup_t |v(x,t)|\right\|_{L^2}.$$

Indeed, if $u \in H$ then

$$\langle\Delta, u\rangle_H = \int_{-\infty}^{\infty} h(x)b(x)dx,$$

where

$$h = \int_0^\infty [(P_tu_t)(Q_tv_t) + (Q_tu_t)(P_tv_t)]\frac{dt}{t} = D\int_0^\infty (P_tu_t)(P_tv_t)dt.$$

By Theorem 4.48, $\|h\|_{L^1} \leq c\|u\|_H \, \|\sup_t |v_t|\|_{L^2}$ which completes the proof.

2.5. The Proof of David and Murai. The method of Calderón pertains, of course, to perturbation theory. The boundedness of T_λ follows from the fact that it is close to the Hilbert transform. David (1984) and Murai (1983) have given yet another perturbation theory variant, allowing them to remove the restriction $M < M_0$ in Calderón's theorem. Until quite recently – before the appearance of the Jones-Semmes proof in Sect. 2.4 of Chapter 4 – this was the simplest route to the boundedness of the Cauchy integral and to the commutator theorem. Murai and Tchamitchian (1984) obtained in this way the best estimate of the norm in terms of M.

We now give the plan of the original argument of Murai (1983).

Consider the operator

$$T = T[\varphi] \stackrel{\text{def}}{=} T[e^{it}, \varphi]$$

with the kernel

$$K[\varphi](x,y) = \frac{1}{x-y}\exp\left\{i\frac{\varphi(x)-\varphi(y)}{x-y}\right\}.$$

It suffices to prove the estimate (5.14), that is the inequality

$$\|T[\varphi]\| \leq c(1+M)^m. \tag{5.21}$$

As we have seen, from this follows the commutator theorem and, in particular, the boundedness of Cauchy integral on an arbitrary Lipschitz curve. Set

$$\tau(M) = \sup\{\|T[\varphi]\| : \|\varphi'\|_{L^\infty} \leq M\}.$$

We know (for instance, by Calderón's Theorem 5.6) that $\tau(M) < \infty$ for $M < M_0$. The transition from the Cauchy integral to the operator $T[\varphi]$ is described in Sect. 2.1.

Theorem 5.11. *$\tau(M) \leq A\tau(\frac{2}{3}M) + BM + C$ where $0 < A, B, C < \infty$.*

Corollary. *$\tau(M) \leq c(1+M)^m$ for some $m < \infty$.*

In order to prove Theorem 5.11 one compares the operator $T[\varphi]$ with the operator $T[\psi]$, where ψ is chosen such that $\|\psi'\|_{L^\infty} \leq \frac{2}{3}M$ and such that, at the same time, φ and ψ differ from each other on a sufficiently big set only by a linear term of the form $ax+b$, which amounts to multiplication of $T[\varphi]$ only by the factor e^{ia} and, in particular, does not change the norm $T[\varphi]$.

Lemma 5.12. *Assume that $\|\varphi'\|_{L^\infty} \leq M$ and let $I \subset \mathbb{R}$ be an interval. There exists a Lipschitz function ψ_I such that $\|\psi_I'\|_{L^\infty} \leq \frac{2}{3}M$ and such that on a closed set $E \subset I$, $|E| \geq \frac{1}{4}|I|$, ψ differs from φ only by a linear term.*

Indeed, let $I = (a, b)$; we may assume that $\varphi(a) = 0$ and $\varphi(b) \geq 0$. Set

$$\begin{aligned}\theta(x) &= \varphi(x) + \frac{1}{3}M(x-a),\\ \theta^*(x) &= \sup\{\theta(y) : a \leq y \leq x\},\ x \in I.\end{aligned}$$

The function θ^* is Lipschitz and $0 \leq \theta^{*\prime} \leq \frac{4}{3}M$. By the "sunrise in the mountains lemma" of F. Riesz (see Riesz and Szökefalvi-Nagy (1972)), the set

$$E = \{x \in I : \theta^*(x) = \theta(x)\}$$

is closed, and θ^* is constant on the contiguous intervals.

Thus,

$$\varphi(b) \leq \theta^*(b) - \frac{1}{3}M|I| \leq \int_E \theta^{*\prime}(x)dx - \frac{1}{3}M|I| \leq \frac{4}{3}M|E| - \frac{1}{3}M|I|,$$

whence $|E| \geq \frac{1}{4}|I|$.

It remains to put

$$\psi_I(x) = \theta^*(x) - \frac{2}{3}M(x-a)$$

and to extend ψ to $\mathbb{R}$ without enlarging the Lipschitz constant.

In order to prove Theorem 5.11, we have to establish an estimate for the distribution function of the form

$$\operatorname{mes}\{x : T_*f(x) > \frac{21}{20}\lambda,\ Mf(x) \le \gamma\lambda\} \le \frac{7}{8}\operatorname{mes}\{x : T_*f(x) > \lambda\} \qquad (5.22)$$

for all $\lambda > 0$. In Sect. 5 of Chapter 1 we have seen that then

$$\|T\| \le c/\gamma. \qquad (5.23)$$

The set $\{T_*f > \lambda\}$ is open. Let I be one of its contiguous intervals, pick a point $\xi \in I$ such that $Mf(\xi) \le \gamma\lambda$, and let ψ_I be the function defined in Lemma 5.12.

As in Sect. 6 of Chapter 3 we write f as a sum of two terms:

$$f = f\chi_{2I} + f\chi_{\mathbf{R}\setminus 2I} = f_1 + f_2.$$

Then $\|f_1\|_{L^1} \le 2\gamma\lambda|I|$ and on I holds

$$T_*f_2(x) \le \lambda + c\gamma\lambda M.$$

Take

$$\gamma \le \frac{1}{40cM}.$$

Then we have $T_*f_2 \le \lambda + \frac{1}{40}\lambda$ on I. We have to show that

$$\operatorname{mes}\{x \in I : T_*f_1(x) > \frac{\lambda}{40}\} \le \frac{7}{8}|I|,$$

which will imply (5.22). As $\varphi(x) = \psi_I(x) + ax + b$ on $E \subset I$, we have for $x \in I$

$$|K[\varphi](x,y) - e^{ia}K[\psi_I](x,y)| \le \frac{8}{3}M\frac{\rho(y,E)}{|x-y|^2},$$

so that on E

$$|T_*f_1(x) - T[\psi_I]_*f_1(x)| \le \frac{8}{3}M\int_{2I}\frac{\rho(y,E)}{|x-y|^2}|f(y)|dy,$$

whence (cf. Sect. 5.5 of Chapter 3)

$$\int_E |T_*f_1(x) - T[\psi_I]_*f_1(x)|dx \le \frac{32M}{3}\int_{2I}|f| \le 22\gamma M\lambda|I|,$$

and

$$\operatorname{mes}\left\{x \in E : T_*f_1 > \frac{\lambda}{40}\right\} \le 180\gamma M|I| + \operatorname{mes}\left\{x \in E : T[\psi_I]_*f_1 > \frac{\lambda}{80}\right\}.$$

As $|E| \geq \frac{1}{4}|I|$ then

$$\text{mes}\left\{x \in I : T_* f_1 > \frac{\lambda}{40}\right\} \leq \frac{3}{4}|I| + 180\gamma M|I| + \text{mes}\left\{x \in E : T[\psi_I]_* f_1 > \frac{\lambda}{80}\right\}$$

Choosing

$$\gamma \leq \frac{1}{16 \cdot 180 \cdot M},$$

we then have to check that

$$\text{mes}\left\{x \in E : T[\psi_I]_* f_1 > \frac{\lambda}{80}\right\} \leq \frac{1}{16}|I|.$$

But $T[\psi_I]$ is a Calderón-Zygmund operator with L^2-norm at most $\tau(\frac{2}{3}M)$ and the constant $M+1$ in the estimates for the kernel. By the weak type estimate (Sect. 6 of Chapter 3)

$$\begin{aligned}\text{mes}\left\{x : T[\psi_I]_* f_1 > \tfrac{\lambda}{40}\right\} &\leq c\left[M + 1 + \tau(\tfrac{2}{3}M)\right] \frac{\|f_1\|_{L^1}}{\lambda} \\ &\leq c\gamma\left[M + 1 + \tau(\tfrac{2}{3}M)\right]|I|.\end{aligned}$$

The desired inequality now follows if we take

$$\gamma \leq \frac{1}{32c[M + 1 + \tau(\frac{2}{3}M)]}.$$

Joining all restrictions on γ, Theorem 5.11 now follows from the inequality (5.23).

The above proof gives a strongly excessive estimate of the norm of $T[\varphi]$, that is, a much too large value of the exponent m in (5.22). Perfecting this, Murai and Tschamitchian (1984) managed to get the following sharp estimate.

Theorem 5.13. *If $\|\varphi'\|_{L^\infty} \leq M$ then the operator*

$$Tf(x) = \text{P.V.} \int_{-\infty}^{\infty} \frac{1}{x - y + i[\varphi(x) - \varphi(y)]} f(y)dy$$

is bounded in $L^2(\mathbb{R}^n)$ and

$$\|T\| \leq c(M + 1)^{1/2}.$$

Recall that the Cauchy integral on the curve $\Gamma = \{x + i\varphi(x), x \in \mathbb{R}\}$ can be viewed as an operator Q in $L^2(\Gamma)$ with respect to arc length and that the norms $\|T\|$ and $\|Q\|$ are connected by the inequalities (5.16).

Corollary. *If Γ is a Lipschitz curve with Lipschitz constant M then the Cauchy integral Q on Γ is bounded in $L^2(\Gamma)$ and*

$$\|Q\| \leq c(M+1)^{3/2}.$$

Note that the proof of Jones and Semmes gives only $\|Q\| \leq c(M+1)^2$. It is though conceivable that it is possible to improve upon this.

David (1986) constructed a Lipschitz curve with Lipschitz constant M for which $\|Q\| \geq c_2(M+1)^{3/2}$, where M can be picked arbitrarily. This means that the estimate of Theorem 5.13 is sharp. In the following section we shall discuss David's counterexample in detail.

Remark. In the proof of Theorem 5.11 one can use the $T1$-Theorem of David and Journé (Theorem 5.2), and estimate $\|T1\|_{\mathrm{BMO}}$ instead of $\|T\|$. As before, Lemma 5.12 plays a key rôle but the proof simplies somewhat.

§3. The Cauchy Integral on Carleson Curves

3.1. Carleson Curves and the Theorem of David. Already in Sect. 6.2 of Chapter 2 we described the solution of the classical problem of the L^p-boundedness of the Cauchy integral on rectifiable Jordan curves.

A rectifiable Jordan curve Γ in the plane is said to be a *Carleson curve* if for every disk $B(z,r)$ holds

$$l(\Gamma \cap B(z,r)) \leq \kappa r \tag{5.24}$$

The *singular integral of Cauchy* is defined by the formula

$$Q_\Gamma f(z) = \lim_{\varepsilon \to +0} \int_{|\zeta - z| > \varepsilon} f(\zeta) \frac{d\zeta}{\zeta - z}. \tag{5.25}$$

The existence of the integral (5.25) and related issues were discussed in detail in Sect. 6 of Chapter 2.

The following definitive result is due to David (1984).

Theorem 5.14. *The operator Q_Γ is bounded in $L^2(\Gamma)$ if and only if Γ is a Carleson curve.*

That condition (5.24) is necessary is easy to see; it suffices to take for f the characteristic function of an arc on Γ. On the other hand, Calderón-Zygmund theory, which in Chapter 3 was developed for operators in $L^p(\mathbb{R}^n)$, extends without any change to the operator Q in $L^p(\Gamma)$. In particular, the maximal operator Q_* (its definition is obvious, sup instead of lim in (5.25)) is likewise bounded in $L^2(\Gamma)$, and Q is bounded in $L^p(\Gamma)$, $1 < p < \infty$, and of weak type $(1,1)$ etc. The weighted estimates in Sect. 7.1 of Chapter 3 carry over to Carleson curves in the following formulation.

Theorem 5.15. *The operator Q_Γ is bounded in $L^p_w(\Gamma)$, $1 < p < \infty$, where w is a weight on Γ if and only if*

$$\sup_{z,r}\left(\frac{1}{r}\int_{B(z,r)} w(\zeta)\,|d\zeta|\right)\left(\frac{1}{r}\int_{B(z,r)} w(\zeta)^{-\frac{1}{p-1}}\,|d\zeta|\right)^{p-1} < +\infty.$$

The same statement remains in force also for the maximal operator Q_*.

The proof of Theorem 5.14 is based on the boundedness of the Cauchy integral in L^2 on a Lipschitz curve with a norm estimate in terms of the Lipschitz constant only. It is a combination of two bold ideas in connection with the technique of Chapter 3. These two ideas are the approximation of Calderón curves with Lipschitz curves and a dual version of the Carleson imbedding theorem.

The complete details of David's proof can be found in David (1986), (1987); here we set forth the main stages only.

3.2. Approximation by Lipschitz Curves

Theorem 5.16. *Let Γ be a Carleson curve and $I \subset \Gamma$ any arc. There exists a Lipschitz curve $\tilde{\Gamma}$ with Lipschitz constant $M = 2\kappa$ such that*

$$l(\tilde{\Gamma} \cap I) \geq \nu l(I),$$

where $\nu = 1/\kappa$ and κ is the constant in condition (5.24).

We give just the construction of the curve $\tilde{\Gamma}$. We may assume that the endpoints of the arc I are at the points 0 and $b \geq 1/\kappa l(I)$. Let $z(s) = x(s)+iy(s)$ be the natural parametrization of I, so that the arc I corresponds to the interval $[0, l(I)]$. If the function $x(s)$ were monotone and had a Lipschitz inverse, then Γ itself would be Lipschitz. Assume that this is not the case. Consider the function (cf. Sect. 2.5)

$$g(s) = x(s) - \frac{b}{2l(I)}s,$$

$$h(s) = \sup\{g(t) : 0 \leq t \leq s\}, \quad 0 \leq s \leq l(I).$$

Finally, set

$$q(s) = h(s) + \frac{b}{2l(I)}s.$$

The function $q(s)$ is increasing and $q'(s) \geq b/2l(I)$. Let φ be the inverse function to q. Then

$$\|\varphi'\|_{L^\infty} \leq 2l(I)/b \leq 2\kappa.$$

The curve with the equation

$$\tilde{z}(s) = x(s) + iy[\varphi(x(s))]$$

is the sought one.

3.3. Duality. Let Γ be a Carleson curve and let $\tilde{\Gamma}$ be Lipschitz. The Cauchy integral $Q_{\tilde{\Gamma}}f$ defines an analytic function off $\tilde{\Gamma}$. Because $\tilde{\Gamma}$ is Lipschitz, this analytic function is in E^2 on all components of $\mathbb{C}\setminus\tilde{\Gamma}$. But in view of the Carleson imbedding theorem (or, more exactly, in view of its analogue for Lipschitz curves, cf. Sect. 2 of Chapter 4)

$$\int_\Gamma |Q_{\tilde{\Gamma}}f|^2 \leq c(\kappa)\|Q_{\tilde{\Gamma}}f\|^2_{E^2} \leq c(\kappa)\int_{\tilde{\Gamma}} |f|^2,$$

where c depends only on κ and the Lipschitz constant of $\tilde{\Gamma}$. In other words, $Q_{\tilde{\Gamma}}$ is a bounded operator from $L^2(\tilde{\Gamma})$ into $L^2(\Gamma)$. Then also the adjoint operator from $L^2(\Gamma)$ into $L^2(\tilde{\Gamma})$ is bounded. The latter, clearly, coincides with Q_Γ. Therefore, even if we do not know whether Q_Γ is bounded on $L^2(\Gamma)$, it is automatically bounded as an operator from $L^2(\Gamma)$ into $L^2(\tilde{\Gamma})$!

With the aid of the technique in Chapter 3 this argument can be carried over to the maximal operator Q_*, which leads to the following result.

Lemma 5.17. *The maximal operator*

$$Q_*f(z) = \sup_{\varepsilon>0}\left|\int_{|\zeta-z|>\varepsilon} f(\zeta)\frac{d\zeta}{\zeta - z}\right|, \quad z \in \tilde{\Gamma},$$

is bounded, as operator from $L^2(\Gamma)$ into $L^2(\tilde{\Gamma})$, and is of weak type as operator from $L^1(\Gamma)$ into $L^1(\tilde{\Gamma})$. Its norm can be estimated in terms of κ and the Lipschitz constant of $\tilde{\Gamma}$ only.

3.4. Estimate of the Distribution Function. Corresponding to the function $f \in L^2(\Gamma)$ let us consider the *maximal function* $Mf(z)$, $z \in \Gamma$, with respect to arc length on Γ. It is defined by the formula

$$Mf(z) = \sup_{r>0}\frac{1}{r}\int_{B(z,r)\cap\Gamma} |f(\zeta)|\,|d\zeta|.$$

Lemma 5.18. *Let $A > 1$, $\nu = 1/\kappa$. Then for $\gamma > 0$ sufficiently small*

$$l\{z \in \Gamma : Q_*f(z) > A\lambda, Mf(z) \leq \gamma\lambda\} \leq (1 - \tfrac{\nu}{2})l\{z \in \Gamma : Q_*f(z) > \lambda\},$$

where $f \in L^2(\Gamma)$, $\lambda > 0$.

If we choose A such that $A^2(1 - \nu/2) < 1$ then, by Sect. 5 of Chapter 1, this gives Theorem 5.14.

The proof of Lemma 5.18 is a repetition of the argument in Sect. 6.3 of Chapter 3. Let I be any of the open arcs which make up the set $\{Q_*f(z) > \lambda\}$ and pick a point $\xi \in I$ such that $Mf(\xi) \leq \gamma\lambda$. Let us show that

$$l\{z \in I : Q_*f(z) > A\lambda\} \leq \left(1 - \frac{\nu}{2}\right) l(I)$$

if γ is sufficiently small.

Let $B = B(\xi, 2l(I))$ and write

$$f = f\chi_B + f\chi_{\Gamma\setminus B} = f_1 + f_2.$$

As always, we obtain $\|f_1\|_{L^1(\Gamma)} \le 2\gamma\lambda l(I)$ and

$$Q_* f_2(z) \le \lambda + c\gamma\lambda, \quad z \in I,$$

with c depending only on Γ. Choose

$$\gamma \le \frac{A-1}{2c}$$

and set

$$E = \left\{ z \in I : Q_* f_1(z) > \frac{A-1}{2}\lambda \right\}.$$

It is sufficient to show that

$$l(E) \le \left(1 - \frac{\nu}{2}\right) l(I).$$

By Theorem 5.16 there exists a Lipschitz curve $\tilde{\Gamma}$ such that $l(\tilde{\Gamma} \cap I) \ge \nu l(I)$, while by Theorem 5.17 the operator Q_* is of weak type $(1,1)$ as an operator from $L^1(\Gamma)$ into $L^1(\tilde{\Gamma})$. In particular

$$l\left\{ z \in \tilde{\Gamma} \cap I : Q_* f_1(z) > \frac{A-1}{2}\lambda \right\} \le \frac{c(\kappa)}{(A-1)\lambda} \|f_1\|_{L^1} \le 2\frac{c(\kappa)}{A-1}\gamma l(I).$$

If we choose

$$\gamma \le \frac{\nu \cdot (A-1)}{4c(\kappa)},$$

then $l(E \cap \tilde{\Gamma}) \le \nu/2 \cdot l(I)$. Hence

$$l(E) \le l(E \cap \tilde{\Gamma}) + l(I \setminus \tilde{\Gamma}) \le \frac{\nu}{2} l(I) + (1-\nu) l(I) = \left(1 - \frac{\nu}{2}\right) l(I).$$

The proof of Theorem 5.14 is complete.

3.5. Multivariate Analogue of David's Theorem. Recently David (1987) obtained an n-dimensional analogue of his theorem, involving k-dimensional surfaces in n-dimensional space.

A rectifiable curve in the plane is given by its natural equation $z = z(s) : \mathbb{R}^1 \to \mathbb{R}^2$.

Let now $z : \mathbb{R}^k \to \mathbb{R}^n$ be a mapping enjoying the following properties:

(i) z is Lipschitz, that is

$$|z(x) - z(y)| \le M|x - y|, \quad x, y \in \mathbb{R}^k.$$

(ii) For any ball $B(w,r) \subset \mathbb{R}^n$

$$|\{x \in \mathbb{R}^k : z(x) \in B\}| \leq \kappa r^k.$$

Such a map z as well as its image $\Gamma = z(\mathbb{R}^k)$ will be referred to as a k-dimensional *Carleson surface* in $\mathbb{R}^n$. On Γ one has two natural measures: the k-dimensional Hausdorff measure and the image under z of Lebesgue measure in $\mathbb{R}^k$. This last measure is defined by the formula

$$\sigma(E) = |\{x \in \mathbb{R}^k : z(x) \in E\}|, \quad E \subset \mathbb{R}^k,$$

or

$$\int_{\mathbb{R}^n} f d\sigma = \int_{\mathbb{R}^k} f[z(x)]dx.$$

The Hausdorff measure is absolutely continuous with respect to the measure σ and the corresponding density is bounded from above and from below. Therefore the space $L^1(\Gamma)$ can be defined used any of these measures and the corresponding norms are equivalent.

Let now $K : \mathbb{R}^n \backslash \{0\} \to \mathbb{C}$ be an infinitely differentiable odd function, homogeneous of degree $-k$, that is

$$K(tx) = t^{-k}K(x), \quad t > 0,\, x \in \mathbb{R}^n.$$

If $f \in L^2(\Gamma)$ set

$$T_* f(z) = \sup_{\varepsilon>0} \left| \int_{|w-z|>\varepsilon} K(z-w)f(w)d\sigma(w) \right|, \quad z \in \mathbb{R}^n.$$

Theorem 5.19. *If Γ is a Carleson surface, then T_* is a bounded operator on $L^p(\Gamma)$, $1 < p < \infty$.*

The condition on the parametrization z of Γ is rather unconvenient. David (1987) gave the following generalization of Theorem 5.19. Let w be a weight in $\mathbb{R}^k$ satisfying the (A_∞) condition of Muckenhoupt (Sect. 4.1 of Chapter 1). Let us assume that, instead of (i) and (ii), z satisfies the conditions

(i′) $|\nabla z| \leq c w^{1/k}$ in distribution sense

and

(ii′) if σ is the image of the measure $w(x)dx$, then $\sigma(B(x,r)) \leq cr^k$ for every ball $B(x,r) \subset \mathbb{R}^n$.

It turns out that on Γ the measure σ is as before equivalent to k dimensional Hausdorff measure and that Theorem 5.19 remains in force for such surfaces.

3.6. The Cantor Set. One might expect that the boundedness in $L^p(\Gamma)$ of the Cauchy integral is only connected with the fact that 1 dimensional

Hausdorff measure on Γ is a Carleson measure, and not with its structure as Jordan curve. Unfortunately, this is not the case. Let us consider the 1 dimensional *Cantor set* $X \subset \mathbb{R}^2$ constructed as follows. Set $E_0 = [0,1]$, $E_0 = [0,1/4] \cup [3/4,1]$ etc. The set E_n consists of 2^n intervals of length 4^{-n}. Set $X_n = E_n \times E_n \subset \mathbb{R}^2$. X_n is the union of 4^n squares $\{Q_k^n\}_{k=1}^{4^n}$ each of side 4^{-n}. Finally, set $X = \bigcap_n X_n$. Then set X is indeed a Cantor set of dimension 1 in $\mathbb{R}^2$.

One dimensional Hausdorff measure on X coincides with the direct product of the two Cantor measures on $E = \bigcap_n E_n$ (concerning Cantor measures see, for example, Gelbaum and Olmsted (1964)). Denoting this measure by μ, we have

$$c_1 r \leq \mu(B(x,r)) \leq c_2 r$$

for any $z \in X$ and $0 < r < 1$.

For $f \in L^2(\mu)$ we define the analogue of the maximal Cauchy integral

$$Q_* f(z) = \sup_{\varepsilon > 0} |Q_\varepsilon f(z)|,$$

$$Q_\varepsilon f(z) = \int_{|\zeta - z| > \varepsilon} f(\zeta) \frac{1}{\zeta - z} d\mu(\zeta).$$

The following result is due to David (1986).

Theorem 5.20. $\|Q_{4^{-n}}1\|_{L^2(\mu)} \geq c\sqrt{n}$, *so that* $Q_*1 \notin L^2(\mu)$ *and the operator* Q_* *cannot be bounded in* $L^2(\mu)$.

To prove this David considers the contribution to the integral $Q_{4^{-n}}1$ of squares of various size

$$f_k(z) = \sum_{Q_l^k \not\ni z} \int_{Q_l^k} \frac{d\mu(\zeta)}{\zeta - z}.$$

Then for $z \in X$

$$(Q_{4^{-n}}1)(z) = \sum_{k=0}^{n} f_k(z).$$

Rather tedious but entirely elementary calculations reveal that $\|f_k\|_{L^2(\mu)} =$ const, while the inner products $\langle f_k, f_l \rangle_{L^2(\mu)}$ drop off exponentially. This yields the estimate $\|Q_{4^{-n}}1\| \geq c\sqrt{n}$.

Modifying this construction and approximating the Cantor set used by an infinite broken line (it runs according to special rules through the corners of the squares Q_k^n), David managed to construct a Lipschitz curve for which the Cauchy integral has a large norm (David (1986)).

Theorem 5.21. *For each* $M > 1$ *there exists a Lipschitz function* f, $\|\varphi'\|_{L^\infty} \leq M$, *such that the norm of the operator*

$$Tf(x) = \text{P.V.} \int_{-\infty}^{\infty} f(y) \frac{1}{x - y + i[\varphi(x) - \varphi(y)]} dy$$

in $L^2(\mathbb{R})$ *satisfies the estimate* $\|T\| \geq 1/10\sqrt{M}$.

Thus, Murai's estimate (Theorem 5.13) for the norm of T is sharp.

Remarks. 1) As is seen from the proof of Jones and Semmes (Sect. 2.4 of Chapter 4), the Cauchy integral defines a bounded operator in $L^2(\Gamma)$ for any Jordan curve Γ such that analytic functions in the components G_+ and G_- of the complement of Γ satisfy the estimate

$$\int_\Gamma |f(z)|^2 |dz| \leq c \iint_{G_\pm} |f'(z)|^2 \rho(z, \Gamma) dx dy. \tag{5.26}$$

The proof of (5.26) mentioned in Sect. 2 of Chapter 4 extends not only to Lipschitz curves but also to arbitrary curves such that the ratio between the length of an arc and the length of the chorde is bounded. However, simple examples ($f(z) = 1/z$ in the domain with boundary $y = \sqrt[4]{|x|}$) reveal that (5.26) does not hold for general Carleson curves. Maybe one has to replace in (5.26) $\rho(z, \Gamma)$ by some other weight, which coincides with $\rho(z, \Gamma)$ in the Lipschitz case. One does not know whether there exists a Littlewood-Paley theory as an adequate tool in the study of the Cauchy integral and also whether it might be possible to prove David's theorem "in one step", without perturbation theory.

On the other hand, the variant of the proof given in Sect. 2 of Chapter 4 gives a worse estimate of the norm of the Cauchy integral on Lipschitz curves (M^2 in place of $M^{3/2}$). It is unlikely that this could be carried over to the general case without considerable changes.

2) If the maximal Cauchy integral Q_* is bounded in L^2 then, invoking weak compactness, one can define the singular integral $Q = \lim Q_\varepsilon$, where the limit is taken over some subsequence $\varepsilon_n \to 0$ in the weak operator topology.

It may be that the unboundedness of Q_* for a Cantor set presents the nonexistence of a "natural and reasonable" regularization Q.

Of course, the question of the boundedness of Q_* is just the most straightforward generalization of the classical problem. It is not known what behavior should be expected from the analytic Cauchy potential near a Cantor set. In analogy with Remark 1 we make the following conjecture.

Conjecture. If $f \in L^2(\mu)$ on X and

$$F(z) = \int_X f(\zeta) \frac{1}{\zeta - z} d\mu(\zeta), \quad z \notin X,$$

then

$$\iint_{\mathbb{C}\setminus X} |F'(z)|^2 \rho(z, X) dx dy < \infty.$$

If this conjecture were true, then it would be clear that David's counterexample shows that the problem is not well posed, and not that the method breaks down.

The considerations of Remark 1 and 2 suggest the following question. Clearly

$$F'(z) = \int_X f(\zeta) \frac{1}{(\zeta - z)^2} d\mu(\zeta).$$

But a singular integral with kernel $(\zeta - z)^{-2}$ is the standard example of a Calderón-Zygmund operator in the plane. Moreover, the operator is applied not to a function in $L^p(\mathbb{R}^2)$, but to the distribution $fd\mu$, and we are interested in the question whether its image is in L^2 with respect to the weight $\rho(z, X)$. An analogous question can be raised in connection with any Calderón-Zygmund operator. We are thus lead to consider nonclassical weighted estimates for such operators.

§4. Supplements

4.1. The Hilbert Transform Once More. In Sect. 7 of Chapter 2 we have remarked that the boundedness of the Hilbert transform

$$Hf(x) = \text{P.V.} \frac{1}{\pi} \int_{-\infty}^{\infty} \frac{1}{t - x} f(t) dt$$

in the Hölder classes $\Lambda^\alpha(\mathbb{R})$, $0 < \alpha < 1$, is much easier to prove than L^2-boundedness. Recently it has been remarked (Meyer (1985), Wittman (1987)) that the L^2-boundedness can be derived from the boundedness in Λ^α with the aid of interpolation of operators.

The idea is the following: as H is a skewsymmetric operator it follows from the boundedness in Λ^α that H is bounded also in $[\Lambda^\alpha]'$, and L^2 is an interpolation space between Λ^α and $[\Lambda^\alpha]'$. But to carry out this program literally leads to considerable technical difficulties. Below we sketch the proof in Wittman (1987).

Lemma 5.22. *Let H be a Hilbert space and $B \subset H$ a Banach space, dense and continuously imbedded in H. Let $T : B \to B$ be an operator such that*

$$\langle Tf, g \rangle_H = -\langle Tg, f \rangle_H, \quad f, g \in B.$$

Then T extends to a bounded operator in H and

$$\|T\|_H \leq \|T\|_B.$$

Indeed, if $f \in B$ and $\|f\|_H = 1$, then

$$\|T^n f\|_H^2 = \langle T^n f, T^n f \rangle = \pm \langle T^{2n} f, f \rangle \leq \|T^{2n} f\|_H,$$

whence

$$\|Tf\|_H \leq \varlimsup_{n \to \infty} \|T^{2^n} f\|^{\frac{1}{2^n}} \leq \varlimsup_{n \to \infty} (c\|f\|_B \|T\|_B^{2^n})^{\frac{1}{2^n}} = \|T\|_B.$$

where c is the norm of the imbedding $B \subset H$.

Let now $\varphi \in \mathcal{D}(\mathbb{R})$, $\varphi(t) = 1$ for $|t| \le 1$, $\varphi(t) = 0$ for $|t| \ge 2$, and set $\varphi_N(t) = \varphi(t/N)$. For any $N > 1$ let us consider the spaces

$$H = L^2(-2N, 2N),$$

$$B = \{f \in \Lambda^\alpha(\mathbb{R}) : f(x) = 0 \text{ for } |x| \ge 2N\},$$

with the Λ^α norm, and the operator

$$T_N f = \varphi_N \cdot H(\varphi_N f)$$

with kernel

$$\frac{1}{\pi}\frac{\varphi_N(x)\varphi_N(y)}{x-y}.$$

Lemma 5.23. $\|T_N\|_B < C < \infty$.

Of course, from this follows, by Lemma 5.22, the boundedness of T_N in $L^2(-2N, 2N)$, that is, the boundedness of H in $L^2(\mathbb{R})$.

But an elementary computation shows that φ_N is a multiplier in B (it sufficces to observe that $\|f\|_{L^\infty} \le cN^\alpha\|f\|_{\Lambda^\alpha}$), so that $g \overset{\text{def}}{=} H(\varphi_N f) \in \Lambda^\alpha$. It is clear that $|g(x)| \le 2\|f\|_{L^\infty} \le cN^\alpha\|f\|_{\Lambda^\alpha}$ for $|x| > 3N$, so that also throughout $\|g\|_{L^\infty} \le cN^\alpha\|f\|_{\Lambda^\alpha}$.

Remark. The proof of Lemma 5.22 is in spirit close to the proof of the Cotlar-Stein Lemma 3.6, so this new approach really does not carry to far away from the previous path.

4.2. Singular Integral Operators in Spaces of Smooth Functions. Let T be a singular integral operator in $\mathbb{R}^n$ in the sense of Sect. 1.1, $T \in \mathrm{SIO}_\alpha$, $0 < \alpha \le 1$.

In what assumptions is it true that T is bounded in the norm of $\Lambda^s(\mathbb{R}^n)$, that is, when do we have

$$\|Tf\|_{\Lambda^s(\mathbb{R}^n)} \le c\|f\|_{\Lambda^s(\mathbb{R}^n)}, \quad f \in \mathcal{D}(\mathbb{R}^n)?$$

Theorem 5.24. *Let $0 < s < \alpha \le 1$ and $T \in \mathrm{SIO}_\alpha$. The operator T is bounded in the Λ^s norm if and only if T is weakly bounded and $T1 = 0$.*

Let us give a sketch of the sufficiency part of Theorem 5.24 (Meyer (1985)). Assume that T be weakly bounded, $T1 = 0$, and let $f \in \mathcal{D}$. Let $x, y \in \mathbb{R}^n$ with $|x - y| = \delta$. Let B be a ball of radius δ containing x and y. Consider a function $\xi \in \mathcal{D}(\mathbb{R}^n)$ which equals 1 in the ball $2B$ and 0 off the ball $4B$, with $|\partial^\beta\xi| \le c_\beta\delta^{-|\beta|}$ for all β. One can show that from $T1 = 0$ follows the identity

$$\begin{aligned} Tf(x) - Tf(y) = {} & T\{\xi(f - f(x))\}(x) - T\{\xi(f - f(y))\}(y) \\ & + \int_{\mathbb{R}^n} [K(x,t) - K(y,t)]\eta(t)[f(t) - f(x)]dt \\ & + [f(x) - f(y)]T\xi(y), \quad \eta = 1 - \xi. \end{aligned}$$

By the estimates on the kernel the two first terms do not exceed

$$c\|f\|_{\Lambda^s}\int_0^{2\delta}\frac{r^s r^{n-1}}{r^n}dr \le c\|f\|_{\Lambda^s}\delta^s,$$

while the third, by the same token, does not exceed

$$c\|f\|_{\Lambda^s}\delta^\alpha\int_\delta^\infty\frac{r^s r^{n-1}}{r^{n+\alpha}}dr \le c\|f\|_{\Lambda^s}\delta^s.$$

Now the boundedness of T in Λ^s follows from the following lemma.

Lemma 5.25. *$|T\xi(x)| \le C < \infty$ for $x \in B$.*

Indeed, as $T1 = 0$ then for $x, y \in B$

$$|T\xi(x) - T\xi(y)| = |T\eta(x) - T\eta(y)| \le c\delta^\alpha\int_0^{2\delta}\frac{r^{n-1}}{r^{n+\alpha}}dr \le c < \infty.$$

Now if $\varphi \in \mathcal{D}$, $\operatorname{supp}\varphi \in B$ and $|\partial^\beta\varphi| \le c_\beta\delta^{-|\beta|}$, then by the weak boundedness of T

$$\left|\int_B T\xi(x)\cdot\varphi(x)dx\right| \le c < \infty.$$

It remains to take $0 \le \varphi \le 1$, $\int_B\varphi \ge \frac{1}{2}|B|$.

Theorem 5.24 extends also to the case $s \ge 1$ (then additional assumptions on T are needed) and to the boundedness in the norm of the Sobolev space $W_2^s(\mathbb{R}^n)$.

A detailed proof can be found in Meyer (1985). In analogy to Sect. 4.1, interpolating between W_2^s and $W_2^{-s} = [W_2^s]^*$ one can derive from these results the $T1$-Theorem of David and Journé (Theorem 5.2), but such an interpolation depends heavily of the Fourier transform and is, unfortunately, more technical than the proof of the $T1$-Theorem itself.

4.3. Singular Integral Operators in BMO

Theorem 5.26. *Let T be Calderón-Zygmund in $\mathbb{R}^n$ and $T1 = 0$. Then T is bounded in* BMO$(\mathbb{R}^n)$.

Indeed, let $f \in \mathrm{BMO}$ and let $B \subset \mathbb{R}^n$ be a ball. Then

$$Tf = T(f - f_{2B}) = T[(f - f_{2B})\chi_{2B}] + T[(f - f_{2B})\chi_{\mathbb{R}^n\setminus 2B}] = Tf_1 + Tf_2.$$

Also

$$\left(\frac{1}{|B|}\int_B |Tf_1|^2\right)^{1/2} \le \|T\|\left(\frac{1}{|B|}\int_B |f - f_{2B}|^2\right)^{1/2} \le c\|T\|\,\|f\|_{\mathrm{BMO}}.$$

On the other hand, putting

$$c_0 = Tf_2(x_0),$$

where x_0 is the center of the ball B, we get for $x \in B$

$$|Tf_2(x) - c_0| \le c \int_{\mathbb{R}^n \setminus 2B} \frac{|x - x_0|^\alpha}{|y - x_0|^{n+\alpha}} |f(y) - f_{2B}| dy \le c\|f\|_{\mathrm{BMO}}$$

(cf. Sect. 9.1 in Chapter 1).

Corollary. *If T is a Calderón-Zygmund operator in $\mathbb{R}^n$ and $T^*1 = 0$ then T is bounded in $H^1(\mathbb{R}^n)$.*

Theorem 5.27. *Let $T \in \mathrm{SIO}_\alpha$ in $\mathbb{R}^n$. The following conditions are equivalent.*
(i) *T is bounded in L^2;*
(ii) *T acts boundedly from L^∞ into BMO;*
(iii) *T acts boundedly from $H^1(\mathbb{R}^n)$ into L^1.*

The proof of Theorem 5.27 can be found in Journé's book (1983). We know already from Chapter 3 that a Calderón-Zygmund operator maps L^∞ into BMO and H^1 into L^1. But the proof of the converse requires a quite elaborate technique.

Sometimes Theorem 5.27 gives better estimates, because it is easier to estimate $\|Tf\|_{\mathrm{BMO}}$ when f is bounded than to get L^2 estimates.

4.4. Spaces of Homogeneous Type. Spaces of homogeneous type have often been mentioned in this part. Although it is not possible to develop their theory here, let us at least give the definition and some references to the literature.

A *space of homogeneous type* is a set X endowed with a pseudo-metric $\rho \ge 0$ and a positive measure μ such that
(i) $\rho(x, y) = \rho(y, x)$;
(ii) $\rho(x, y) = 0$ if and only if $x = y$;
(iii) $\rho(x, y) \le M[\rho(x, z) + \rho(z, y)]$;
(iv) all balls $B(x, r) = \{y \in X : \rho(y, x) < r\}$ are measurable and

$$\mu(B(x, 2r)) \le \kappa\mu(B(x, r)), \quad x \in X,\ r > 0.$$

On spaces of homogeneous type one can define in a natural way the spaces L^p and Λ^s.

A detailed discussion of the theory of spaces of homogeneous type can be found in Calderón (1976), Coifman and Weiss (1971), (1977), Folland and Stein (1982), Macias and Segovia (1979a), (1979c), Strömberg and Torchinsky (1980), Uchiyama (1980).

One can carry over the maximal theorem and the weighted estimates to such spaces, and likewise a considerable part of Calderón-Zygmund theory.

The Littlewood-Paley theory in Coifman's formulation and the proof of the Tb-Theorem also can be extended to these spaces.

The following are examples of spaces of homogeneous type:

1) Any Lipschitz manifold in $\mathbb{R}^n$ with the Hausdorff measure.

2) Homogeneous groups in the sense of Folland and Stein (1982); in particular, the Heisenberg group.

3) Carleson surfaces in $\mathbb{R}^n$ in the sense of Sect. 3.5.

4) Any compact subset of $\mathbb{R}^n$ with the homogeneous measure constructed in Vol'berg and Konyagin (1984).

5) $\mathbb{R}^n$ with the usual distance and the measure $w(x)dx$ where w satisfies the Muckenhoupt condition (A_∞).

6) $\mathbb{R}^n$ with Lebesgue measure but with a parabolic distance (cf. Sect. 2.6 of Chapter 3).

Annotated Literature

Quadratic expressions as a tool for estimating L_p-norms were introduced in Analysis by Littlewood-Paley (1931/36). The subsequent development of Littlewood-Paley theory is reflected in the books Zygmund (1968), Stein (1970a), (1970b), Folland and Stein (1982) and in the surveys Coifman and Weiss (1978), Cowling (1981) and Stein (1982).

The far best introduction to the theory is Chapter IV of Stein's book (1970a) and, what concerns more recent developments, the book Stein and Folland (1982).

The conference proceedings (1979), (1982), (1983) and (1985) illustrate the current state of affairs and the most recent trends of development of the subject.

The paper David, Journé, and Semmes (1985) (or (1986)) is a model for the interplay between Littlewood-Paley theory and estimates for singular integral operators, continuing the tradition of the work of Fefferman and Stein (1972) and Coifman, Meyer, and Stein (1983), (1985).

We refer the reader to two new text books on the theory of singular integrals: Garcia-Cuerva, and Rubio de Francia (1985) and Journé (1983). For Littlewood-Paley theory of martingales and the interplay between harmonic analysis and probability theory see Gikhman, and Skorohod (1982), Burkholder (1979a), (1979b), Durrett (1984), Garcia (1973), Gundy (1980), Varopoulos (1980).

A brilliant exposition of the history and the theory of wavelets can be found in Meyer (1987). Meyer gives a long list of forrunners of wavelets, to which we may add the atomary functions of V. A. Rvachev (1986).

For the application of Calderón commutators to boundary problems for differential equations we refer to Calderón (1980), Dahlberg and Kenig (1985), Fabes, Jerison, and Kenig (1982), Fabes, Jodeit, and Lewis (1977), Fabes, Jodeit, and Rivière (1978), Verchota (1984). Dahlberg, and Kenig (1985) contains an extensive bibliography.

The literature for the $T1$- and Tb-Theorems was given in the text. The attempts for a direct extension of the theory of the Cauchy integral to the multidimensional case (except for the method of rotations) have so far not been very successful. For one of several possible variants see Brackx, Delange, and Sommer (1982).

Concerning applications of Littlewood-Paley theory to spaces of smooth functions and imbedding theorems we may quote, besides Stein (1970a), the books Maz'ya (1985), Peetre (1976) and Triebel (1978), (1983).

For the theory of spaces of homogeneous type see the book Folland, and Stein (1982) and also the articles Vol'berg, and Konyagin (1984), Calderón (1977), Calderón and Torchinsky (1975/77), Coifman, and Weiss (1971), (1977), Macias, and Segovia (1979a), (1979b), (1979c), Strömberg, and Torchinsky (1980), Uchiyama (1980).

Bibliography*

Aguilera, N. and Segovia, C. (1977): Weighted norm inequalities relating the g_λ^* and the area functions. Stud. Math. *61*, No. 3, 293–303, Zbl. 375.28015.

Battle, G. and Federbush, P. (1982): A phase cell cluster expansion for Euclidean field theories. Ann. Phys. *142*, 95–139.

Belinskiĭ, P. P. (1974): General properties of quasiconformal mappings. Moscow: Nauka (98 pp.) [Russian] Zbl. 292.30019.

Brackx, F., Delange, R., and Sommer, F. (1982): Clifford analysis. New York: Pitman, Zbl. 529.30001.

Burkholder, D. L. (1979a): Martingale theory and harmonic analysis on Euclidean spaces. In: Harmonic analysis on Euclidean spaces, Proc. Symp. Pure Math. *35*, part II, 283–301. Providence, RI: Am. Math. Soc., Zbl. 417.60055.

Burkholder, D. L. (1979b): A sharp inequality for martingale transforms. Ann. Probab. *7*, No. 4, 858–863, Zbl. 416.60047.

Burkholder, D. L. and Gundy, R. F. (1972): Distribution function inequalities for the area integral. Stud. Math. *44*, No. 6, 527–544, Zbl. 219,179.

Calderón, A. P. (1965): Commutators of singular integral operators. Proc. Natl. Acad. Sci. USA *53*, 1092–1099, Zbl. 151,169.

Calderón, A. P. (1976): Inequalities for the maximal function relative to a metric. Stud. Math. *57*, 297–306, Zbl. 341.44007.

Calderón, A. P. (1977): Cauchy integrals on Lipschitz curves and related operators. Proc. Natl. Acad. Sci. USA *74*, 1324–1327, Zbl. 373.44003.

Calderón, A. P. (1980): Commutators, singular integrals on Lipschitz curves and applications. In: Proc. Int. Congr. Math., Helsinki 1978, vol. 1, 85–96, Zbl. 429.35077.

Calderón, A. P. and Torchinsky, A. (1975/77): Parabolic maximal functions associated with a distribution. Adv. Math. *16*, 1–64, Zbl. 315.46037; *24*, 101–171, Zbl. 355.46021.

Carleson, L. (1967): Selected problems on exceptional sets. Princeton: Van Nostrand. (98 pp.), Zbl. 189,109.

Chang, S. Y. A. and Fefferman, R. (1980): A continuous version of the duality of H^1 with BMO on the bidisk. Ann. Math., II. Ser. *112*, 179–201, Zbl. 451.42014.

Chang, S. Y. A. and Fefferman, R. (1985): Some recent developments in Fourier analysis and H^p theory on product domains. Bull. Am. Math. Soc., New Ser. *12*, 1–43, Zbl. 557.42007.

Coifman, R., David, G., and Meyer, Y. (1983): La solution des conjectures de Calderón. Adv. Math. *48*, No. 2, 144–148, Zbl. 518.42024.

Coifman, R., McIntosh, A., and Meyer, Y. (1982): L'intégrale de Cauchy définit un opérateur borné sur L^2 pour les courbes Lipschitziennes. Ann. Math., II. Ser. *116*, No. 2, 361–387, Zbl. 497.42012.

* For the convenience of the reader, references to reviews in Zentralblatt für Mathematik (Zbl.), compiled using the MATH database, and Jahrbuch über die Fortschritte der Mathematik (Jbuch) have, as far as possible, been included in this bibliography.

Coifman, R. and Meyer, Y. (1975): On commutators on singular integrals and bilinear singular integrals. Trans. Am. Math. Soc. *212*, 315–331, Zbl. 324.44005.

Coifman, R. and Meyer, Y. (1978): Au-delà des opérateurs pseudo-differentiels. Asterisque *57* (188 pp.), Zbl. 483.35082.

Coifman, R., Meyer, Y., and Stein, E. M. (1983): Un nouvel espace fonctionnel adapté à l'étude des opérateurs définis par des intégrales singulières. In: Harmonic Analysis, Proc. Conf. Cortona/Italy 1982, Lect. Notes Math. *992*, 1–15, Zbl. 523.42016.

Coifman, R., Meyer, Y., and Stein, E. M. (1985): Some new function spaces and their applications to harmonic analysis. J. Funct. Anal. *62*, 304–335, Zbl. 569.42016.

Coifman, R. and Weiss, G. (1971): Analyse harmonique non-commutative sur certains espaces homogénes. Lect. Notes Math. *242*. (160 pp.), Zbl. 224.43006.

Coifman, R. and Weiss, G. (1977): Extensions of Hardy spaces and their use in analysis. Bull. Am. Math. Soc. *83*, 569–645, Zbl. 358.30023.

Coifman, R. and Weiss, G. (1978): Book review of "Littlewood-Paley and multiplier theory" by R. E. Edwards and G. I. Gaudry. Bull. Am. Math. Soc. *84*, 242–250.

Cowling, M. G. (1981): On Littlewood-Paley-Stein theory. Rend. Circ. Mat. Palermo, II. Ser., Suppl. 1, 21–55, Zbl. 472.42012.

Dahlberg, B. (1977): Estimates of harmonic measure. Arch. Ration. Mech. Anal. *65*, 275–288, Zbl. 406.28009.

Dahlberg, B. (1980): Weighted norm inequalities for the Luzin area integrals and the non-tangential maximal functions for functions harmonic in a Lipschitz domain. Stud. Math. *67*, 297–314, Zbl. 449.31002.

Dahlberg, B. and Kenig, C. (1985): Hardy spaces and the Neumann problem in L^p for Laplace equation in Lipschitz domains I. Ann. Math., II. Ser. *125*, No. 3, 437–465, Zbl. 658.35027.

Daubechies, I., Grossmann, A., and Meyer, Y. (1986): Painless non-orthogonal expansions. J. Math. Phys. *27*, 1271–1283, Zbl. 608.46014.

David, G. (1984): Opérateurs intégraux singuliers sur certains courbes du plan complexe. Ann. Sci. Ec. Norm. Super., IV. Ser. *17*, No. 1, 157–189, Zbl. 537.42016.

David, G. (1986): Noyau de Cauchy et opérateurs de Calderón-Zygmund. Thèses d'Etat, Univ. Paris-Sud (210 pp.).

David, G. (1987): Opérateurs d'intégrales singulières sur les surfaces regulières. Preprint, École Polytechnique (49 pp.). Appeared in: Ann. Sci. Ec. Norm. Super., IV Ser. *21*, No. 2, 225–258, (1988), Zbl. 655.42013.

David, G. and Journé, J.-L. (1984): A boundedness criterion for generalized Calderón-Zygmund operators. Ann. Math., II. Ser. *120*, No. 2, 371–397, Zbl. 567.47025.

David, G., Journé, J.-L., and Semmes, S. (1985): Opérateurs de Calderón-Zygmund, fonctions para-accretive et interpolation. Rev. Mat. Iberoam. *1*, No. 4, 1–56, Zbl. 604.42014.

Durrett, R. (1984): Brownian motion and martingales in analysis. Belmont, California: Wadsworth (328 pp.), Zbl. 554.60075.

Dyn'kin, E. M. (1987): Methods of the theory of singular integrals (Hilbert transform and Calderón-Zygmund theory). In: Itogi Nauki Tekh, Ser. Sovrem. Probl. Mat. *15*, 197–202. Moscow: VINITI. English translation: Encyclopaedia Math. Sci. Vol. 15, pp. 167–259. Berlin Heidelberg: Springer-Verlag (1991), Zbl. 661.42009.

Dyn'kin, E. M. (1981): A constructive characterization of Sobolev and Besov classes. Tr. Mat. Inst. Steklova *155*, 41–76. English translation: Proc. Steklov Inst. Math. *155*, 39–74 (1983), Zbl. 496.46021.

Fabes, E., Jerison, D., and Kenig, C. (1982): Multilinear Littlewood-Paley estimates with applications to partial differential equations. Proc. Natl. Acad. Sci. USA *79*, 5746–5750, Zbl. 501.35014.

Fabes, E., Jodeit, M. jun., and Lewis, T. (1977): Double layer potentials for domains with corners and edges. Indiana Univ. Math. J. *26*, 95–114, Zbl. 363.35010.

Fabes, E., Jodeit, M. jun., and Rivière, N. (1978): Potential theoretic techniques for boundary value problems on C^1-domains. Acta Math. *141*, 165–186, Zbl. 402.31009.

Fefferman, Ch. (1972): The multiplier problem for the ball. Ann. Math., II. Ser. *94*, 330–336, Zbl. 234.42009.
Fefferman, Ch. and Stein, E. M. (1972): H^p spaces of several variables. Acta Math. *129*, 137–193, Zbl. 257.46078.
Folland, G. B. and Stein, E. M. (1982), Hardy spaces on homogeneous groups. (Mathematical Notes 28.) Princeton: Univ. Press (284 pp.), Zbl. 508.42025.
Garnett, J. B. (1981): Bounded analytic functions. New York: Academic Press (467 pp.), Zbl. 469.30024.
Garsia, A. (1973): Martingale inequalities. Reading, Mass.: Benjamin (184 pp.), Zbl. 284.60046.
Garcia-Cuerva, J. L. and Rubio de Francia, J. (1985): Weigthed norm inequalities and related topics. Amsterdam New York Oxford: North-Holland (604 pp.), Zbl. 578.46046.
Gelbaum, B. R. and Olmsted, I. M. (1964): Counterexamples in analysis. San Francisco, CA: Holden-Day (194 pp.), Zbl. 121,289.
Gikhman, I. I. and Skorokhod, A. V. (1982): Stochastic differential equations and their applications. Kiev: Naukova Dumka. [Russian], Zbl. 557.60041.
Goluzin, G. M. (1966): Geometric theory of functions of a complex variable. Moscow: Nauka (628 pp.). English translation: Translation of Mathematical Monographs 26. Providence, RI: Amer. Math. Soc. (1969), Zbl. 148,306.
Gundy, R. F. (1980): Inégalités pour martingales à un et deux indices: l'espace H^p. In: Ecole d'été de probabilites de Saint-Fleur VIII – 1978, Lect. Notes Math. *774*, 251–334, Zbl. 427.60046.
Hunt, R. and Wheeden, R. (1968): On the boundary values of harmonic functions. Trans. Am. Math. Soc. *132*, No. 2, 307–322, Zbl. 159,405.
Jerison, D. and Kenig, C. (1980): An identity with applications to harmonic measure. Bull. Am. Math. Soc., New Ser. *2*, No. 3, 447–451, Zbl. 436.31002.
Jerison, D. and Kenig, C. (1982a): Boundary behavior of harmonic functions in nontangentially accesible domains. Adv. Math. *46*, No. 3, 80–147, Zbl. 514.31003.
Jerison, D. and Kenig, C. (1982b): Hardy spaces, $(A)_\infty$ and singular integrals on chord arc domains. Math. Scand. *50*, 221–247, Zbl. 509.30025.
Journé, J.-L. (1983): Calderón-Zygmund operators, pseudo-differential operators and the Cauchy integral of Calderón. Lect. Notes Math. *994*, (128 pp.), Zbl. 508.42021.
Koosis, P. (1980): Introduction to H^p spaces. Cambridge: Univ. Press (376 pp.), Zbl. 435.30001.
Kurtz, D. (1980): Littlewood-Paley and multiplier theory on weighted L^p spaces. Trans. Am. Math. Soc. *259*, No. 1, 235–254, Zbl. 436.42012.
Larsen, R. (1971): An introduction to the theory of multipliers. (Grundlehren Math. Wiss. 175). Berlin Heidelberg: Springer-Verlag (282 pp.), Zbl. 213,133.
Lemarié, P. G. and Meyer, Y. (1986): Ondelettes et bases hilbertiennes. Rev. Mat. Iberoam. *2*, 1–18, Zbl. 657.42028.
Littlewood, J. E. and Paley, R. E. A. C. (1931/36): Theorems on Fourier series and power series. J. Lond. Math. Soc. *6*, 230–233, Zbl. 2,188; Proc. Lond. Math. Soc., II. ser. *42*, 52–89, Zbl. 15,254; *43*, 105–126, Zbl. 16,301.
Macias, R. and Segovia, C. (1979a): Lipschitz functions on spaces of homogeneous type. Adv. Math. *33*, 257–270, Zbl. 431.46018.
Macias, R. and Segovia, C. (1979b): A decomposition into atoms of distributions on spaces of homogeneous type. Adv. Math. *33*, 271–309, Zbl. 431.46019.
Macias, R. and Segovia, C. (1979c): Singular integrals on generalized Lipschitz and Hardy spaces. Stud. Math. *65*, No. 1, 55–75, Zbl. 479.42014.
McIntosh, A. and Meyer, Y. (1985): Algèbres d'opérateurs définis par les intégrales singulières. C. R. Acad. Sci., Paris, Ser. I *301*, No. 8, 395–397, Zbl. 584.47030.
Maz'ya, V. G. (1985): Sobolev spaces. Leningrad: LGU (416 pp.). English translation: Berlin Heidelberg: Springer-Verlag (1985), Zbl. 692.46023.

Meyer, Y. (1985): Continuité sur les espaces de Hölder et de Sobolev des opérateurs définis par les intégrales singulières. In: Recent progress in Fourier analysis, Proc. Semin., El Escorial/Spain 1983, North-Holland Math. Stud. *111*, 145–172, Zbl. 616.42008.
Meyer, Y. (1987): Wavelets and operators. Preprint CEREMADE N° 8704, Univ. Paris-Dauphine (108 pp.). Appeared in: Lond. Math. Soc. Lect. Note Ser. *137*, 256–365 (1989).
Muckenhoupt, B. and Wheeden, R. (1974): Norm inequalities for the Littlewood-Paley function g_λ^*. Trans. Am. Math. Soc. *191*, 95–111, Zbl. 289.44005.
Murai, T. (1983): Boundedness of singular integrals of Calderón type. Proc. Japan Acad., Ser. A *59*, No. 8, 364–367, Zbl. 542.42008.
Murai, T. and Tschamitchian, P. (1984): Boundedness of singular integral operators of Calderón type V, VI. Preprint series, College of general education, Nagoya, No. 8; No. 12. Appeared in: Adv. Math. *59*, 71–81, Zbl. 608.42010.
Peetre, J. (1976): New thoughts on Besov spaces. (Duke Univ. Math. Series 1.) Durham, N. C.: Math. Department, Duke Univ. (304 pp.), Zbl. 356.46038.
Pommerenke, Ch. (1978): Schlichte Funktionen und analytische Funktionen von beschränkter mittlerer Oszillation. Comment. Math. Helv. *52*, No. 4, 591–602, Zbl. 369.30012.
Privalov, I. I. (1950): Boundary values of analytic functions. Moscow Leningrad: GITTL (336 pp.), Zbl. 41,397.
Riesz, F. and Sz.-Nagy, B. (1972): Leçons d'analyse fonctionelle. Budapest: Akad. Kiadó English translation: New York: Frederick Ungar 1978 (467 pp.), Zbl. 46,331; Zbl. 64,354.
Rudin, W. (1973): Functional analysis. New York: McGraw-Hill (432 pp.), Zbl. 253.46001.
Rvachev, V. A. (1986): Atomary functions and their applications. In: The theory of *R*-functions and contemporary problems of applied mathematics, pp. 45–65. Kiev: Naukova Dumka [Russian].
Stein, E. M. (1970a): Singular integrals and differentiability properties of functions. Princeton: Univ. Press (290 pp.), Zbl. 207,135.
Stein, E. M. (1970b): Topics in harmonic analysis related to the Littlewood-Paley theory. Princeton: Univ. Press (146 pp.), Zbl. 193,105.
Stein, E. M. (1982): The development of square functions in the work of A. Zygmund. Bull. Am. Math. Soc., New Ser. *7*, 359–376, Zbl. 526.01021.
Strichartz, R. S. (1967): Multipliers on fractional Sobolev spaces. J. Math. Mech. *16*, 1031–1060, Zbl. 145,383.
Strömberg, J.-O. and Torchinsky, A. (1980): Weights, sharp maximal functions and Hardy spaces. Bull. Am. Math. Soc., New Ser. *3*, 1053–1056, Zbl. 452.43004.
Triebel, H. (1978): Interpolation theory, function spaces, differential operators. Berlin: VEB Wiss. Verlag (528 pp.) and North-Holland Publ. Co., Zbl. 387.46032.
Triebel, H. (1983): Theory of function spaces. Basel: Birkhäuser (432 pp.), Zbl. 546.46027.
Uchiyama, A. (1980): A maximal function characterization of H^p on the space of homogeneous type. Trans. Am. Math. Soc. *262*, 579–592, Zbl. 503.46020.
Varopoulos, N. T. (1980): Aspects of probabilistic Littlewood-Paley theory. J. Funct. Anal. *38*, No. 1, 25–60, Zbl. 462.60050.
Verchota, G. (1984): Layer potentials and regularity for the Dirichlet problem for Laplace's operator in Lipschitz domains. J. Funct. Anal. *59*, No. 3, 572–611, Zbl. 589.31005.
Vol'berg, A. P. and Konyagin, S. V. (1984): On every compact set in $\mathbb{R}^n$ there exists a homogeneous measure. Dokl. Akad. Nauk SSSR *278*, No. 3, 783–786. English translation: Sov. Math., Dokl. *30*, 453–456 (1984), Zbl. 598.28010.
Wittmann, R. (1987): Application of a theorem of M. G. Kreĭn to singular integrals. Trans. Am. Math. Soc. *299*, No. 2, 581–599, Zbl. 596.42005.
Zygmund, A. (1959): Trigonometric series I–II. Cambridge: Univ. Press (383 pp.; 354 pp.), Zbl. 85,56.
Zygmund, A. (1979): Harmonic analysis on Euclidean spaces, Part I–II, Proc. Symp. Pure Math. *35*. Providence, RI: Amer. Math. Soc. (898 pp.), Zbl. 407.00005, Zbl. 407.00006.
Zygmund, A. (1982): Harmonic analysis, Proc. Minneapolis, 1981, Lect. Notes Math. *908*. Berlin: Springer-Verlag (326 pp.), Zbl. 471.00014.

Zygmund, A. (1983): Harmonic analysis. Proc. Conf. Cortona/Italy, 1982, Lect. Notes Math. *992*. Berlin: Springer-Verlag (450 pp.), Zbl. 504.00013.
Zygmund, A. (1985): Recent progress in Fourier analysis (eds. I. Peral and J. L. Rubio de Francia). Amsterdam: North-Holland (268 pp.), Zbl. 581.00009.

III. Exceptional Sets in Harmonic Analysis

S. V. Kislyakov

Translated from the Russian
by J. Peetre

Contents

Introduction

"Exceptional" (or "thin") sets are often encountered in works on harmonic analysis. Many deep investigations are devoted to them and many outstanding unsolved problems are connected with them. So also the division "Commutative Harmonic Analysis" of this series has not been able to do without an article especially devoted to them.

Nevertheless there exists no discipline which could be called the "theory of exceptional sets". Exceptional sets are rather heterogeneous what concerns their origin and their nature. Also the methods and techniques used to study various classes of exceptional sets differ strongly from each other. In one word, even for the specialist it is difficult to answer what are precisely the common features of all objects collected together under this heading (though they undoubtedly have some such features).

The author also did not aim at the task of describing this unity in proper mathematical terms. Therefore, Sect. 1 of the present survey may be viewed simply as an incomplete catalogue of systems of exceptional sets, and the reader is asked not to consider too seriously the general arguments and explanations this catalogue is accompanied with. The remaining two sections are devoted to some concrete classes of exceptional sets. In Sect. 2 some rather recent developments in the theory of Sidon sets are presented, while Sect. 3 contains material in some way or other connected with the notion of capacity. The choice of precisely these two themes out of the immense sea of possibilities was dictated by the following considerations. First, from the methodological point of view these two themes may be viewed as "extremal poles" in the theory of exceptional sets (which, we reiterate, does not exist). Second, things connected with capacities belong, undoubtedly, to the most important part of the theory, while Sidon sets is a domain where progress during the last years has been particularly spectacular. Lastly, most of the other themes connected with exceptional sets will be (or have already been) treated in other parts of this series.

Trying to evade criticism from the part of the reader, we wish to emphasize that a more satisfactory illumination of the history of the questions touched upon in Sects. 2, 3 is hardly possible in a work of limited size, and is not at all connected with the author's possible disinterest in such matters.

I am grateful to V. P. Khavin for valuable critical remarks.

§1. Main Constructions

1.1. Unmathematical Definition. Intituitively, exceptional set are sets so "thin" that on them one encounters a behavior which is not displayed by ordinary ("big") sets. As a rule, it is about a special behavior on the sets under consideration of appropriate elements of some function space[1]. To each such behavior corresponds, of course, an appropriate system of exceptional sets. We attempt, however, to give a general descriptive definition of such systems, independent of the process how it is constructed.

Let G be a locally compact Abelian group. A family $\mathcal{F}$ of closed subsets of G is called a *system of exceptional sets* if the following conditions are fulfilled.

(i) If $A \subset B \in \mathcal{F}$ and A is closed, then $A \in \mathcal{F}$.

(ii) If $A \in \mathcal{F}$ and $g \in G$ then $A + g \in \mathcal{F}$.

(iii) All sets in $\mathcal{F}$ are "small" in some sense.

This definition serves only as a crude basis for the following discussion, and we will not always pedantically adhere even to what is entirely strict there. Thus, the requirement that all exceptional sets should be closed is often far too heavy (but, of course, it is always necessary to restrict the class of sets under consideration, by requiring, for instance, that they should be Borelian or analytic etc.).

From a not very formal point of view condition (iii), appealing to the intuition, may be taken to coincide with the precise condition (i) – because (i) says that the size of a set is more decisive for the set to be in $\mathcal{F}$ than its structure. However, in practice one speaks of a class $\mathcal{F}$ subject to the conditions (i) and (ii) as a class of exceptional sets only if some additional property of its members can be singled out (regardless of which character it may have, topological, arithmetic, metric etc.), which intuitively is perceived as a "condition of smallness".

Let us consider one very general method to define exceptional sets. Let X be a locally convex space of functions on the group G. For each closed set E, $E \subset G$, denote by $X(E)$ the set of all restrictions of functions in X to E. Let us assume that the space X is translation invariant, and let Y be any other locally convex translation invariant space of functions on G. Then the class $\mathcal{F}$, $\{E \subset G : E \,\text{closed and}\, X(E) = Y(E)\}$, satisfies the conditions (i) and (ii).

We shall now see that in very general assumptions $\mathcal{F}$ consists of sets which in a precise (although crude) sense are "small". As usual $A(G) = \{\hat{f} : f \in L^1(G)\}$.

Lemma 1.1.1. *If X and Y are modules over the algebra $A(G)$ with respect to pointwise multiplication, $X \neq Y$ and G is compact, then all sets in $\mathcal{F}$ have empty interior.*

Proof. If this were not the case, there would exist an open set U such that the sets of restrictions $X|_U$ and $Y|_U$ coincide. Every translate of U has the same property and a finite number of translates covers G. Let $x_1, \ldots, x_n$ be

[1] This term is taken in a wide sense: the space may consist of, say, measures, pseudomeasures, distributions ...

a partition of unity subordinated to this finite cover and such that $x_i \in A(G)$ for each i. If $f \in X \cup Y$, then $f = x_1 f + \cdots + x_n f$ with all $x_i f$ in $X \cap Y$. Thus $X = Y$.

Let us now check that for the group of integers $\mathbb{Z}$ (which we have chosen as the simplest example of an infinite discrete group) the above description likewise often leads to small sets, but in a different sense, of course. We assume that X and Y are $A(\mathbb{Z})$ Banach modules such that in both of them the set of functions with finite support is dense. In addition, we assume that all translations are isometries on both X and Y.

Lemma 1.1.2. *In the assumptions made, let $X \neq Y$. Then $\mathcal{F}$ does not contain sets containing arbitrarily large intervals.*

Proof. The algebra $A(\mathbb{Z})$ contains characteristic functions of all singletons on $\mathbb{Z}$. Consequently, convergence in any of the spaces X and Y implies pointwise convergence. Hence, in view of the closed graph theorem, we see that for $E \in \mathcal{F}$ the spaces $X(E)$ and $Y(E)$ not only coincide but have equivalent norms (we intend here, of course, the quotients of X and Y under the restriction map $f \mapsto f|_E$). Therefore, if the conclusion of the lemma were not true, then for all intervals I the norms in $X(I)$ and $Y(I)$ would be equivalent, with a constant independent of I.

Let v_n be the Fourier transform of the de la Vallée-Poussin kernel: $v_n = V_n|_{\mathbb{Z}}$, where V_n is a continuous function on $\mathbb{R}$, vanishing on the set $\{t : |t| \geq 2n\}$, equal to unity on the interval $[-n, n]$ and linear on $[-2n, -n]$ and on $[n, 2n]$. The functions v_n are uniformly bounded on $A(\mathbb{Z})$. It is easy to see that if $f \in X$ then $v_n f \to f$ in X and the same is true if we replace X by Y. The uniform equivalence of the norms now shows that $X = Y$.

Let $\Gamma_0, \Gamma_1, \ldots, \Gamma_k$ be the cosets of $\mathbb{Z}$ with respect to a fixed subgroup Γ_0. An analogous reasoning shows that if $E \in \mathcal{F}$ then there exist intervals which for no choice whatsoever of $m_0, m_1, \ldots, m_k$ in Γ_0 are contained in the union $\bigcap_{0 \leq j \leq k} [(E \cup \Gamma_j) + m_j]$.

We shall not dwell on further considerations in this direction. In particular, we will not explain in what sense the sets obtained by the above construction for generall locally compact Abelian groups G are "small" (although it would have been possible to do this). Instead, we pass to concrete examples.

1.2. Helson Sets. Sidon Sets. The Helson sets arise if we take as X and Y the spaces $A(G)$ $(= A)$ and $C_0(G)$ $(= C_0)$ in the construction of Sect. 1.1 (the last space consists of the continuous functions on G vanishing at infinity). In other words, a closed set E, $E \subset G$, is said to be a *Helson set*, if every continuous function on E that vanishes at infinity can be extended to a function in $A(G)$.

If E is a Helson set then, by the closed graph theorem, the number $\alpha(E)$,

$$\alpha(E) \stackrel{\text{def}}{=} \sup\{\|f\|_{A(E)} : f \in A(E), \|f\|_{C_0(E)} = 1\},$$

called the *Helson constant* of the set E, is finite. (Of course, also the converse is true: if $\alpha(E) < \infty$ then E is Helson.) The following theorem is one of the main achievements of the theory of Helson sets.

Theorem 1.2.1. *The union of two Helson sets is a Helson set.*

A proof (with the estimate $\alpha(E_1 \cup E_2) \leq 2^{-1}3^{2/3}(\alpha(E_1)^3 + \alpha(E_2)^3))$ can be found in e. g. Graham and McGehee (1979). There one can likewise find the history of the question, and references to the literature, as well as a discussion of other aspects of Helson sets. Some information about them in the case $G = \mathbb{T}$ can be found in the survey Kislyakov (1987) in this series. Cf. furthermore the books Kahane (1968) and Rudin (1962).

In the case when G is a discrete group Helson sets are called *Sidon sets* (and the Helson constant is called the *Sidon constant*). Section 2 will be devoted to Sidon sets. Here we give just another (dual) definition of them. It is of interest also because by the same scheme one can define some other classes of exceptional sets too.

Let G be discrete. If Y is any class of functions on G and if $E \subset G$, then we shall denote by Y_E the set $\{y \in Y : y|_{(G\setminus E)} = 0\}$. Let X be a translation invariant locally convex space of functions on G, which is furthermore a topological module over $A(G)$ with respect to pointwise multiplication. Let us assume that the functions with compact (that is, finite) support are dense in X. Then the dual X^* can be identified with a space of functions on G: to any continuous linear functional F, $F \in X^*$, there corresponds a function φ such that $\varphi(g) = F(\mathbf{1}_{\{g\}})$, $g \in G$. (We shall denote throughout the characteristic function of a set a by $\mathbf{1}_a$.) It is clear that the function φ determines F uniquely. The duality in this realization of X^* is given by

$$\langle x, \varphi \rangle = \sum_{g \in G} x(g)\varphi(g) \tag{1}$$

for all functions x with finite support.

If $E \subset G$, then, as we have already noted, one can identify $X(E)$ with the quotient space $X/X_{G\setminus E}$. Thus, $X(E)^*$ is nothing but the annihilator of $X_{G\setminus E}$ in X^*.

Lemma 1.2.2. *We have* $(X_{G\setminus E})^{\perp} = (X^*)_E$.

Proof. If $\varphi \in (X_{G\setminus E})^{\perp}$, then $\langle \mathbf{1}_{\{g\}}, \varphi \rangle = 0$ for all g in $G\setminus E$, i. e. the function φ vanishes on $G\setminus E$. Conversely, if this statement is true, then, in view of (1), $\langle x, \varphi \rangle = 0$ for every function x with compact support, not intersecting E. It remains to remark that the set of all such functions x is dense in $X_{G\setminus E}$. Indeed, the algebra $A(G)$ contains an approximate identity, i. e. a bounded directed net $\{\lambda_\alpha\}$ such that the support of each λ_α is finite and $\lambda_\alpha(g) \to 1$ for all g, $g \in G$. Under the assumptions imposed on X it is easy to see that $\lambda_\alpha x \to x$ in X for each function x, $x \in X$ (as this, clearly, holds true for every x with finite support). Finally, if $x \in X_{G\setminus E}$, then $\lambda_\alpha x$ has finite support and vanishes on E.

Let Y be another function space satisfying the same assumptions as X. Under suitable auxiliary assumptions one can then conclude that the equations $X(E) = Y(E)$ and $(X^*)_E = (Y^*)_E$ are equivalent. For example, this is so if X and Y both are Banach spaces. Indeed, if $X(E) = Y(E)$ then it follows from the closed graph theorem that this equality is not only set theoretical but also topological, so that $X(E)^* = Y(E)^*$, i. e. $(X^*)_E = (Y^*)_E$. Conversely, if $(X^*)_E = (Y^*)_E$ then this equality clearly is topological and therefore $X(E)$ and $Y(E)$ have equivalent norms on the intersection of these spaces, which again is dense in any of them.

Let us see what this gives for Sidon sets. For a discrete group G the space $A(G)^*$ coincides with the space of Fourier transforms of functions in $L^\infty(\hat{G})$; moreover, $C_0(G)^* = l^1(G)$ (where $l^1(\Lambda)$, as usual, stands for the Banach space of all absolutely summable families of numbers on the set Λ). Thus, the set E, $E \subset G$, is a Sidon set if and only if each function f in $L^\infty(\hat{G})$ with spectrum in E satisfies the condition $\sum_{g\in G} |\hat{f}(g)| < \infty$.

1.3. Sets of Type Λ_p. Let again G be a discrete group. Conditions of the type "$X_E = Y_E$" do not define just Sidon sets but also other classes of exceptional sets. Among these we have in particular the set of type Λ_p. Let $1 \leq q < p < \infty$. Set $X = (L^p(\hat{G}))\hat{}$ and $Y = (L^q(\hat{G}))\hat{}$. Using Hölder's inequality it is not hard to see that the class of those sets E for which $X_E = Y_E$ is independent of q, $q < p$. Such sets are called *sets of type* Λ_p. Equivalent formulation: E is a set of type Λ_p if for some (and, then, every) q, $q < p$, the norms of the space $L^p(\hat{G})$ and $L^q(\hat{G})$ are equivalent on the set of functions on $\hat{G}$ of the form $\sum_{g\in E} a_g g$, with $a_g \neq 0$ only for finitely many characters g.

Besides the literature on sets of type Λ_p indicated in Kislyakov (1987) we mention further Bourgain (1988) and Lopez and Ross (1975). In Bourgain's paper there is constructed, for each p, $p \geq 2$, a set of type Λ_p which is not of type $\Lambda_{p+\varepsilon}$ for any ε, $\varepsilon > 0$.

1.4. What is X_E in the Case of a Nondiscrete Group? Synthesis. Let now G, generally speaking, be nondiscrete, X as before being a translation invariant space of functions on G, which is a (topological) module over $A(G)$, the functions in $A(G)$ with compact support forming a dense subset in E. If $X \subset C(G)$ then one can find functions in X concentrated to E only if E has non-empty interior. If we wish to define X_E for not too "big" sets E, we are forced to consider much bigger spaces X, i. e. we must avoid imposing the condition $X \subset C(G)$.

It follows from the preceding subsection that there are also other natural reasons for not imposing such an inclusion. For example, if we want to identify the space $X(E)^*$ (this is one of the candidates for $(X^*)_E$) then we have to invoke the space X^*, which in the case $X \subset C(G)$ often cannot be identified with a space of functions on G. For instance, if $X = C_0(G)$ then X^* is the space $M(G)$ of bounded Borel measures on G, which can be realized as a space of functions only if G is discrete (then $M(G) = l^1(G)$).

Another important example (where the situation is more complicated than for $C_0(G)$) is the space $A(G)$. The continuous linear functionals on $A(G)$ are called *pseudomeasures*. In contrast to $C_0(G)$, the action of pseudomeasures on Fourier transforms of functions in $A(G)$ is easily described, but not on the functions themselves. Namely, Φ is a pseudomeasure if and only if there exists a function F in $L^\infty(\hat{G})$ such that

$$\Phi(\hat{f}) = \int fF dm, \quad f \in L^1(G), \tag{2}$$

where m is the Haar measure on G.

If G is compact then the dual group $\hat{G}$ is discrete, while $L^1(\hat{G})$ coincides with $l^1(\hat{G})$, that is, with $C_0(\hat{G})^*$. Thus, for a compact group G the space $A(G)$ itself is a dual; its predual is the set of those pseudomeasures Φ for which the function F in the representation (2) belongs to $C_0(\hat{G})$. Such pseudomeasures are termed *pseudofunctions*.

Returnig to the general case, we are faced with the problem of giving a meaning to the words "a linear functional F, $F \in X^*$ vanishing off a closed set E" (or "supported on E"). In analogy with Lemma 1.2.2 we could have said that they are equivalent to the containment $F \in (X_{G \setminus E})^\perp$, where $X_{G\setminus E} \overset{\text{def}}{=} \{x \in X : x|_E = 0\}$. However, in practice one proceeds differently. In what follows we shall assume that convergence in X implies pointwise convergence. Then $X_{G\setminus E}$ is a closed submodule of X. The point is that this is not the simplest of the closed submodules Y satisfying the condition $E = \{g \in G : y(g) = 0 \text{ for all } y \text{ in } Y\}$.

The set $G\setminus E$ is open, and from the conditions imposed on X at the beginning of this subsection it follows that each function in $A(G)$ with compact support not intersecting E belongs to $X_{G\setminus E}$. The closure of the set of all such function in X will be denoted by $X^0_{G\setminus E}$. It is clear that $X^0_{G\setminus E} \subset X_{G\setminus E}$. If equality holds in this inclusion, then we say that E is a *set of (spectral) synthesis* for X. In general, not every set is of this type.

A linear functional Φ, $\Phi \in X^*$, is said to be supported to the set E if $\Phi \perp X^0_{G\setminus E}$. The set of all linear functional supported on E will be denoted $(X^*)_E$. The least closed set on which a linear functional is supported (if there exists such a set) is called its *support.*

Thus we have the equality $X(E)^* = (X^*)_E$ if and only if E is a set of synthesis for X. This circumstance, and the very notation $(X^*)_E$ is not in contradiction with the contents of Sect. 1.2 (in particular, not with Lemma 1.2.2), as each subset of a discrete group is a set of synthesis for every space X (of course, subject to some natural assumptions). Incidently, this was in fact established in the course of the proof of Lemma 1.2.2.

Synthesis holds true for every closed set E if $X = C_0(G)$ (and any group G). A measure μ, $\mu \in M(G) = C_0(G)^*$, is supported on a set E if and only if $|\mu|(G \setminus E) = 0$, so that $(C_0(G)^*)_E = M(E) = C_0(E)^*$. On the contrary, if $X = A(E)$ then for a non-discrete group G there exist always subsets which are not sets of synthesis, cf. e. g. Kahane (1970), Rudin (1962). The problem of spectral synthesis in $A(G)$ is the subject of study of one of the most interesting chapters in harmonic analysis. However, this problems falls off the limits of the question discussed in this part. Let us remark that whether a given set is a set of synthesis or not depends rather on its "structure" than on its "size". Therefore the system of sets of synthesis, as well as the system of sets of nonsynthesis, is not a system of exceptional sets in the sense of Sect. 1.1. On the other hand, one has studied the class of closed sets such that each subset is a set of synthesis for $A(G)$ (such sets are called "*R-sets*"). This class is a system of exceptional sets, provided the group is non-discrete. See, for example, Kahane (1970).

1.5. Continuation of the Catalogue. For a while, let us assume that G is compact. For such groups we defined in the previous section the notion of pseudofunction. Every pseudofunction is a pseudomeasure, so that it is meaningful to speak of its support. A closed set is said to be a *set of uniqueness* if every pseudofunction supported to it vanishes, and a *set of uniqueness in the narrow sense* if every measure μ supported on E and such that $\hat{\mu} \in \check{C}_0(\hat{G})$ vanishes. Both these classes (we will denote them by U and U_0 respectively) are systems of exceptional sets. Within the framework of this series they are taken up for discussion in Kislyakov (1987); there one can likewise find references for further reading.

Let us remark that both these definitions are of the same nature as the "dual" definition of Sidon set in Sect. 1.2: the classes U and U_0 are singled out by an identity of the type $X_E = Y_E$; one simply takes now $Y = \{0\}$ (and X is either the space of pseudomeasures or the space of measures whose Fourier transforms drop off at infinity).

We remark furthermore that, by a well-known theorem by Rajchman (it is given together with its proof in Kislyakov (1987)), the measure μ in the definition of U_0 can be assumed to be nonnegative. By the same scheme (now we do not assume anymore that G is compact) one defines many other important classes of exceptional sets: they are all distinguished by the requirement that on them there does not exist nonnegative measures whose Fourier transform satisfies an appropriate growth condition. If this condition has the form

$$\int_{\hat{G}} |\hat{\mu}(\xi)|^2 \omega(\xi) dm(\xi) < \infty$$

(ω being a positive function of a special type and m is the Haar measure on G), then we get the *sets of capacity zero*. Of course, if ω changes then the last notion changes too. The word "capacity" has also an independent meaning, appearing not only in connection with the term just introduced. We shall discuss the material connected with capacities in detail in Sect. 3.

If $\omega \equiv 1$ then the sets of capacity zero are just subsets of G of zero Haar measure. Let us also remark that for an almost general ω the "verification" of condition (iii) in Sect. 1.1 is not hard, because it turns out that sets of capacity zero by necessity must have empty interior.

In the last example we applied the construction of the type X_E to a set of nonnegative measures, not worrying that this was a nonlinear class. In principle a construction of the type $X(E)$ also can be applied to various nonlinear subsets of linear spaces. Thus *Kronecker sets*, which are very important in harmonic analysis, are defined in this way; in many (but not all) respects they are the "narrowest" exceptional sets. A compact set E, $E \subset G$, is said to be a Kronecker sets if the set of restrictions to E of all of all characters is dense (in the uniform topology) in the set of all continuous unimodular functions on E. (Note that a subset of a Kronecker set is not necessarily a Kronecker set, cf. Lindahl and Paulsen (1971), p. 9, that is, property (i) in the definition in Sect. 1.1 fails. Besides, in finite-dimensional groups such a behavior is not possible, cf. Rudin (1962).)

Each Kronecker set is at the same time a set of synthesis, a Helson set and a set of uniqueness (cf. the table in Kahane (1970), Chapter VII, Sect. 1). If G does not contain elements of finite order (or if there are not "too many" such elements), then G contains a Kronecker set and, moreover, there is one in each perfect set of G. In the case when there exist "too many" elements of finite order, the last statement fails, but it can be "restored", if we modify the notion of Kronecker set in a suitable way (the so-called classes K_q). We will not pursue further discussion of these classes of sets and their applications here, but refer the reader to Hewitt and Ross (1963/70), Kahane (1970) and Rudin (1962), as well as to V. P. Gurariĭ's survey in Vol. 25 of this series.

1.6. Conclusion. We interrupt this catalogue of systems of exceptional sets, without attempting to complete it, or even to approach completeness. Let us sum up what has been said. The previous pages have been guided by the idea that given a set E there exist only two canonical constructions: the construction of $X(E)$, which to each class of functions X assigns the class of restrictions of these functions to E, and the construction X_E, taking into account those objects in X that are supported on E. Moreover, these constructions are intimitely connected with each other, as they are in a sense dual to each other. The question of their complete duality lead to the *problem of spectral synthesis*. Systems of exceptional sets are often distinguished by the condition of coincidence of the results of these constructions when applied to different spaces X and the same set E. The condition of "narrowness" (condition (iii) in Sect. 1.1) in such constructions can often be given a rigorous formulation.

Not denying these facts, let us give just one example, which underlines their uncompleteness. Let A^+ be the space of those functions in $A(\mathbb{T})$ ($\mathbb{T}$ is the circle group) for which $\hat{f}(n) = 0$ for $n < 0$. Set

$$ZA^+ = \{E \subset \mathbb{T} : \exists f, f \in A^+, f \neq 0, f|E = 0\}.$$

This is a system of exceptional sets and their theory is rather extensive (cf., for example, Kahane (1970), Chapter VIII). Nevertheless the definition of the class ZA^+ does not fall into the previous scheme. (Also, one observes that ZA^+ consists of the sets E such that $(A^+)_E \neq 0 \dots$)

§2. Sidon Sets

2.1. Problems. To the two equivalent definitions of Sidon sets in Sect. 1 we shall add yet another one. Let G be a compact Abelian group and let Γ be the discrete group dual to it (usually written multiplicatively).

Definition. A subset E of Γ is said to be a *Sidon set* if for each function Φ in $l^\infty(\Gamma)$ one can find a measure μ in $M(G)$ such that $\hat{\mu}|_E = \Phi$.

We remark that in view of the open mapping principle the definition remains unchanged if we also impose the estimate $\|\mu\| \leq \kappa\|\Phi\|$ for some constant κ independent of μ. The infimum of all such constants κ is called the Sidon constant of E.

This definition is often used when one wants to construct concrete examples of Sidon sets and then the measure μ can be found in the form of a *Riesz product*:

$$\mu = \prod_{\gamma \in E} (1 + \mathrm{Re}\,[\Phi(\gamma)\gamma]). \tag{3}$$

Of course, in the most general case formula (3) does not make sense and we are faced with the problem how to give a sense to it.

We say that a set E, $E \subset \Gamma$, is *strongly independent* if the relation $\prod_{\gamma \in E} \gamma^{\varepsilon(\gamma)} = 1$, where ε is a function mapping E into $\mathbb{Z}$ and $\{\gamma : \varepsilon(\gamma) \neq 0\}$ is a finite set, holds only for $\varepsilon \equiv 0$.

Let us assume that E is strongly independent and that $|\Phi| \leq 1$. Consider finite partial products of the infinite product in (3):

$$g_F = \prod_{\gamma \in F} (1 + \mathrm{Re}\,[\Phi(\gamma)\gamma]), \quad \mathrm{card}\, F < \infty.$$

They are all nonnegative functions whose integral equals 1. This follows from the strong independence of E. From this it is easy to see now that the directed set $\{g_F m\}$ (where m is the Haar measure on G and the sets F are ordered by inclusion) has a unique limit point in the weak $*$-topology on the space $M(G)$. Thus, (3) has been given a meaning. Moreover, it is not hard to see that $\hat{\mu}(\gamma) = 2^{-1}\Phi(\gamma)$ for $\gamma \in E$.

Thus, all strongly independent sets are Sidon sets (with constant ≤ 2). Formula (3) gives also additional information: namely, that the measure μ, which solves the interpolation problem $\hat{\mu}|_E = \Phi$ for a strongly independent set E, can be taken to be nonnegative.

A typical example of an infinite strongly independent set is the family of coordinate functions $\{z_k\}$ on the infinite dimensional torus $\mathbb{T}^\infty$. However, we stress that there are rather few strongly independent sets; for example, on the group $\mathbb{Z}$ all such sets are singletons. Luckily, in the preceding two paragraphs we have not used strongly independent sets to their full extent – for example, we can as easily give meaning to (3) if E satisfies only the following condition: $1 \in E$ and $\prod_{\gamma \in E_0} \gamma_{\varepsilon(\gamma)} = 1$, where E_0, $E_0 \subset E$, is finite and $\varepsilon(\gamma) \in \{-2, -1, 0, 1, 2\}$, implies that $\gamma^{\varepsilon(\gamma)} = 1$ for all γ in E_0. Such a set E is said to be *dissociated*. Moreover, formula (1) makes sense also if $E = F \cup F^{-1}$ where F is a dissociated, but then Φ has to be taken *Hermitean*, that is, such that $\Phi(\gamma^{-1}) = \overline{\Phi(\gamma)}$. (Of course, for E to be Sidon it suffices to solve the problem $\hat{\mu}|_E = \Phi$ for Hermitean functions Φ only.) Let us formulate this as a theorem. The details (not difficult!) can be found in Lopez and Ross (1975), pp. 19–21.

Theorem 2.1.1. *Let $E = F \cup F^{-1}$ where F is a dissociated subset of the group Γ. Then for every Hermitean function Φ on E there exists a nonnegative measure such that $\hat{\mu}|_E = \Phi$ and $\|\mu\| \leq 2\|\Phi\|_\infty$. In particular, F is a Sidon set.*

In the group $\mathbb{Z}$ every set of positive integers $\{n_k\}$ is dissociated if, say, $n_{k+1}/n_k \geq 3$. If we have only $n_{k+1}/n_k \geq \lambda > 1$, then we say that $\{n_k\}$ is *lacunary in the sense of Hadamard.* There is a classical theorem which states that every finite union of sets lacunary in the sense of Hadamard is a Sidon set, cf. for example Zygmund (1959).

A set E, $E \subset F$, is said to be a *Fatou-Zygmund* set, if the conclusion of Theorem 2.1.1 holds true for it: there exists a constant κ, $\kappa > 0$, such that for each Hermitean function Φ on E there exists a nonnegative measure μ such that $\hat{\mu}|_E = \Phi$ and $\|\mu\| \leq \kappa\|\Phi\|_\infty$ (in the definition of Hermiticity it is, of course, sufficient to require the condition $\Phi(\gamma^{-1}) = \overline{\Phi(\gamma)}$ for $\gamma, \gamma^{-1} \in E$ only).

Now we can finally approach the goal of this subsection, namely the formulation of the problems which have ruled the development of the theory of Sidon sets in the past 15–20 years. The results of this development will be illuminated in Sect. 2.2. The first problem pertains to the Fatou-Zygmund sets (throughout the years a great number of papers have been devoted to them; cf. the comments in Lopez and Ross (1975)).

Problem 1. What is the relation between Sidon sets and Fatou-Zygmund sets?

The idea to use Riesz products to prove that a given set is Sidon can be applied also to sets that are not dissociated. Here we mention only one of the basic intermediate steps: it is sufficient to solve the problem $\hat{\mu}|_E = \Phi$ in an "approximative" sense and only for functions Φ that take the values ± 1. How this done is indicated in Lopez and Ross (1975), Chapter 1–2. Let us just state the endresult.

Let $\Lambda \subset \Gamma$, with $1 \neq \Lambda$ and $\gamma \in \Gamma$. For every nonnegative number s denote by $R_s(\gamma, \Lambda)$ the number of families $\{\varepsilon_\lambda\}_{\lambda\in\Lambda}$ with $\varepsilon_\lambda \in \{-1, 0, 1\}$, $\sum_{\lambda\in\Lambda} |\varepsilon_\lambda| = s$ and $\gamma = \prod_{\lambda\in\Lambda} \lambda^{\varepsilon_\lambda}$. Furthermore set $R(\gamma, \Lambda) = \sum_{s\geq 0} R_s(\gamma, \Lambda)$.

A set Λ, $1 \neq \Lambda$, is said to be *quasi-independent* if $R(1, \Lambda) = 1$ (or, what is the same, if $R_s(1, \Lambda) = 0$ for $s \geq 1$), it is called a *Rider set* if $\sum_{s\geq 0} \delta^s R_s(1, \Lambda) < \infty$ for some positive δ [2] and a *Stechkin set* if it is a finite union of Rider sets.

Theorem 2.1.2. (cf. Lopez and Ross (1975), p. 30). *Every Stechkin set is a Sidon set.*

We have stated already in Sect. 1 that a finite union of Sidon sets is a Sidon set. On the other hand, so far one has no examples of Sidon sets besides Stechkin sets. There is a conjecture that Stechkin sets also exhaust all examples. The following two problems are connected with this.

Problem 2. Is it true that every Sidon is a finite union of quasi-independent sets?

Let us remark that every Rider (and, therefore, every Stechkin set) is a finite union of quasi-independent sets; cf. for example Pisier (1983), Proposition 2.13.

Problem 3. Give an arithmetic characterization of Sidon sets.

By an arithmetic characterization we mean, for example, a characterization in terms of families $\{\varepsilon_\lambda\}_{\lambda\in\Lambda}$ such that $\varepsilon_\lambda = 0, \pm 1$, $\sum_{\lambda\in\Lambda} |\varepsilon_\lambda| < \infty$ and $\prod_{\lambda\in\Lambda} \lambda^{\varepsilon_\lambda} = 1$. A possible solution of Problem 2 might also solve Problem 3.

Finally, the last problem is connected with the following result, which has been known for a long time (cf. Lopez and Ross (1975), Chapter 5).

Theorem 2.1.3. *Let E be a Sidon set. Then E is a set of type Λ_p for every p, $1 \leq p < \infty$. Moreover, for each finite linear combination u of characters in E we have the relation* $\|u\|_{L^p(G)} \leq \text{const} \cdot p^{1/2} \|u\|_{L^2(G)}$, $2 < p < \infty$.

Problem 4. Does the last conclusion of Theorem 2.1.3 imply that E is a Sidon set?

---

[2] This condition guarantees the convergence of the appropriate Fourier coefficients of the Riesz product.

Incidently, let us remark that Theorem 2.1.3 allows to obtain various "narrowness conditions" on the Sidon sets, cf. Lopez and Ross (1975), pp. 78–81. Thus, if E is a Sidon set in $\mathbb{Z}$ with Sidon constan κ, A being a finite arithmetic progression in $\mathbb{Z}$, then $\operatorname{card}(E \cap A) \leq 6\kappa e \log \operatorname{card} A$. In the reference quoted one finds likewise a variant of this result for general groups.

Let us give yet another narrowness condition: a Sidon set can not for any infinite sets $A, B \subset \Gamma$ contain their product $A \cdot B$.

2.2. Solutions. To the list of problems in the preceding section we might have added the question whether the union of two Sidon sets is a Sidon set, but the answer to this has already been given in Sect. 1. This answer, and the solution to Problem 1 is connected with the name Drury, who has devised a clever method for averaging "interpolating" measures. The general theorem on the union of Helson set mentioned in Sect. 1 was also inspired by Drury's work (cf. the comments to it in Graham and McGehee (1979)).

Drury's results on Sidon sets were obtained in the early 70's. We summarize them in the following theorem (for the proof consult Lopez and Ross (1975), Chapter 3).

Theorem 2.2.1. *Let E, $E \subset \Gamma$, be a Sidon set and let $0 < \varepsilon < 1$. If $\Phi \in l^\infty(E)$ then there exists a measure μ on G such that $\hat{\mu}|_E = \Phi$, $\|\mu\| \leq C\varepsilon^{-1}$ and $|\hat{\mu}(\gamma)| \leq \varepsilon^{-1}$ for $\gamma \notin E$ (the constant C depends on the Sidon constant of E only). If $1 \notin E$ and Φ is Hermitean, then μ can be taken to be positive.*

Corollary 1. *A Sidon set which does not contain unity is a Fatou-Zygmund set.*

This solves Problem 1.

Corollary 2. *The union of two Sidon sets is a Sidon set.*

This follows readily from the fact that the Fourier transform of a measure μ satisfying the condition $\hat{\mu}|_E = \Phi$ can be made arbitrarily small off E (cf. Theorem 2.2.1). Before Drury's results sets with this propery had been studied specially under the name "uniform Sidon sets".

Further progress in the theory of Sidon sets has been connected with the application of probabilistic considerations. Riders' paper (1975) has been pivotal here. We give its result in Pisier's treatment (1977–78a). Let $\{\varepsilon_k\}$ be independent stochastic variables, each taking only the values $+1$ and -1 with the same probability.

Theorem 2.2.2. *Let $E \subset \Gamma$. Assume that there exist a constant K such that for each finite subset $\{\gamma_1, \ldots, \gamma_n\}$ of E and arbitrary scalars $\alpha_1, \ldots, \alpha_n$ the inequality*

$$\sum_{1 \leq i \leq n} |\alpha_i| \leq K \int \| \sum_{1 \leq i \leq n} \varepsilon_i(n) \alpha_i \gamma_i \|_{C(G)} dP(\omega) \tag{4}$$

holds. Then E is a Sidon set.

The converse of this theorem is trivial, so that condition (4) in fact does characterize Sidon sets. The integral to the right in (4) diminishes if we restrict the set $\{\gamma_1, \dots, \gamma_n\}$, so that Theorem 2.2.2 gives us another proof of the fact that the union of two Sidon sets is Sidon. Also in the proof of Theorem 2.2.2 one uses the ideas of Drury.

Pisier (1977–78a) has remarked that if the random variables ε_k in (4) are replaced by independent equidistributed Gaussian random variables, then the condition so obtained likewise characterizes Sidon sets. This observation (not difficult when Theorem 2.2.2 is known) leads to the idea that there might be a connection between the theory of Sidon sets and the theory of almost surely continuous random (Gaussian) Fourier series. Such a connection has also been uncovered by Pisier, see Marcus and Pisier (1981), Pisier (1977–78b), (1981), (1983). As a result he obtained thereby, in particular, the solution to Problems 3 and 4 in Sect. 2.2.

Let $\{\gamma\}_{\gamma\in\Gamma}$ be a family of independent standard Gaussian stochastic variables. Denote by $C_{\text{p.s.}}(G)$ the space of all functions f in $f \in L^2(G)$ such that the norm $|||\cdot|||$, $|||f||| \stackrel{\text{def}}{=} \sup\{E\|\sum_{\gamma\in A} g_\gamma \hat{f}(\gamma)\gamma\|_\infty\}$, where the least upper bound is taken over all finite subsets of Γ, is finite. The Dadly-Fernique theorem on the continuity of trajectories of the standard Gaussian process applied to the space $C_{\text{p.s.}}(G)$ takes the following form.

For a function f in $L^2(G)$ set $d_f(s,t) = \|f_s - f_t\|_{L^2}$ (here $s, t \in G$ and $f_y(x) = f(x+y)$). The function d_f is a (quasi-)metric on G, and therefore one can consider the entropy $N(G, f, \varepsilon)$, the least number of d_f-balls of radius ε by which one can cover G.

Theorem 2.2.3. *There exists a constant C, independent of f, such that*

$$C^{-1}|||f||| \le |\hat{f}(1)| + \int_0^{2\|f\|_2} (\log N(G, f, \varepsilon))^{1/2} d\varepsilon \le C\|f\|. \qquad (5)$$

For the proof see Marcus and Pisier (1981), Pisier (1977–78b). In Marcus and Pisier (1981) considerably more general results than the ones formulated here are given. (Cf. furthermore Marcus and Pisier (1984), where the full analogue of inequality (5) is obtained for p-stable stochastic variables instead of Gaussian ones. In that paper applications are given to Sidon sets in the spirit set forth below (for more detailed information we refer the reader to Marcus and Pisier (1984), Pisier (1983))).

Theorem 2.2.3 allows us to describe the space $C_{\text{p.s.}}(G)^*$ (for details see Marcus and Pisier (1984), Pisier (1977/78b)). Set $\psi(x) = \exp|x|^2 - 1$ and let $L_\psi(G)$ be the Orlicz space corresponding to the function ψ. We denote by $M(\psi)$ the space of *Fourier multipliers* (i. e. translation invariant operators) from L^2 into L_ψ. Recall that each such multiplier T is determined by its Fourier coefficients $\hat{T}(\gamma)$: $T(\gamma) = \hat{T}(\gamma)\gamma$, $\gamma \in \Gamma$.

Theorem 2.2.4. *$C_{\text{p.s.}}(G)^* = M(\psi)$ in the natural duality $\langle f, T\rangle = \sum \hat{f}(\gamma)\hat{T}(\gamma)$.*

Corollary. *Problem 4 has a positive solution.*

Indeed, assume that the set E satisfies the conclusion of Theorem 2.1.3. Then for each function β in $l^\infty(E)$ the multiplier T defined by the conditions $\hat{T}(\gamma) = 0$ for $\gamma \notin E$, $\hat{T}(\gamma) = \beta(\gamma)$ for $\gamma \in E$, belongs to $M(\psi)$. In this way we get a continuous operator $\beta \mapsto T$ from $l^\infty(E)$ into $M(\psi)$. The continuity of the adjoint operator means, in view of Theorem 2.2.4, that $\sum_{\gamma\in E} |\hat{f}(\gamma)| \leq C\|f\|_{C_{\text{p.s.}}(G)}$ for every function f with spectrum in E. Thus, E is Sidon (cf. the remarks following Theorem 2.2.2).

Let us state two more characteristics of Sidon sets due to Pisier, connected with Theorem 2.2.2 and a positive answer to Problem 4 (cf. Pisier (1981), (1983)).

Theorem 2.2.5. *Let $E \subset \Gamma$. The following conditions are equivalent.*

(i) *E is a Sidon set.*

(ii) *For each finite subset $A \subset E$, holds the estimate $\|\sum_{\gamma\in A} \gamma\|_{L_\psi} \leq C(\operatorname{card} A)^{1/2}$ (C does not depend on A).*

(iii) *There exists a number $\delta > 0$, such that for each finite subset $A \subset E$, the inequality $\|\sum_{\gamma\in A} \varepsilon_\gamma \gamma\|_\infty \geq \delta(\operatorname{card} A)$ holds (here ε_γ has the same meaning as in Theorem 2.2.2).*

Furthermore, Pisier obtained the following result, the core of the proof being again the "entropy" inequality (5).

Theorem 2.2.6. *To the conditions* (i)–(iii) *one can add still another two:*

(iv) *(In the supplementary a priori assumption $1 \notin E$.) There exists a number θ, $0 < \theta < 1$, such that for each finite subset $A \subset E$, holds the inequality*

$$\sum_{0\leq s\leq \operatorname{card} A} 2^{-s} R_s(1, A) \leq 2^{\theta \operatorname{card} A}.$$

(v) *There exists a number $\delta > 0$, such that each finite subset A of E contains a quasi-independent subset B such that* $\operatorname{card} B \geq \delta \operatorname{card} A$.

The equivalence of the statements (i) and (v) is a very strong indication in favor of a positive answer to the question in Problem 2, the only one among the problems in Sect. 2.1 which so far ise not solved. Condition (iv) gives an arithmetic characterization of Sidon sets and, thus, solves Problem 3. For the proof of Theorem 2.2.6 see Pisier (1983), where likewise other arithmetic characterizations can be found.

2.3. Complements. The theory of stochastic processes has, as we have seen, played an important rôle in the theory of Sidon sets. However, it turned out later that part of the results of the last section can be obtained also without them. Namely, Bourgain (1985) has established the equivalence of the following four statements by constructing special Riesz products.

1. E is Sidon set.

2. $\|\sum_{\gamma\in E} a_\gamma \gamma\|_p \le Cp^{1/2}(\sum_{\gamma\in E} |a_\gamma|^2)^{1/2}$, $p \ge 2$.

3. The same as condition (v) in Theorem 2.2.6.

4. There exists a number $\delta > 0$ such that if $\{a_\gamma\}_{\gamma\in E}$ is any finite sequence of scalars then there exits a quasi-independent set A, $A \subset E$, such that $\sum_{\gamma\in A} |a_\gamma| \ge \delta \sum_{\gamma\in E} |a_\gamma|$.

"Randomness" is present also in the above paper by Bourgain, but in a quite elementary form: in order to obtain a Riesz product with the desired properties one establishes that a suitable family of such products has this property "on the average".

Let us state still two results based on "entropy" characterizations of Sidon sets found by Pisier. The first of them is due to Bourgain (1985). Let $M_d(G)$ be the space of all discrete measures on G.

Theorem 2.3.1. *The equality* $\{\hat{\mu}|_E : \mu \in M(G)\} = \{\hat{\mu}|_E : \mu \in M_d(G)\}$ *implies that* E *is a Sidon set.*

The second result, due to Bourgain and Milman (1985) concludes in some sense a theme taken up more than 10 years earlier by Varopoulos. (He proved in Varopoulos (1976) that if the Banach space $C_E(G)$ is linearly homeomorphic to l^1 then E is a Sidon set.)

Theorem 2.3.2. (cf. Bourgain and Milman (1985)). *Assume that in the space* $C_E(G)$ *it is not possible to find a sequence of finite dimensional subspaces* X_n *such that* $\dim X_n = n$ *and* $\sup_n d(X_n, l_n^\infty) < \infty$ *(here* $d(\cdot,\cdot)$ *is the Banach-Mazur distance). Then* E *is Sidon.*

With this theorem we conclude our survey of recent achievements in the theory of Sidon sets. An important guide to material preceding it is the monograph Lopez and Ross (1975). Among the problems illuminated there but not included here we mention the important question when one has the inequality $\sum_{\gamma\in E} |a_\gamma| \le \text{const} \cdot \sup |\sum_{g\in I} a_\gamma \gamma(g)|$ if E is a Sidon set and I an open subset of G. We note however that this problem is almost solved. At least the information found in Lopez and Ross (1975) is rather satisfactory. See further Bourgain (1985).

§3. Capacities and Related Topics

3.1. Potential, Energy, Capacity. Let G be a locally compact Abelian group, k a positive continuous function on $G\backslash\{0\}$, having at the origin the limit $+\infty$ (such a function will be called a *kernel*; we will always define it at the origin by the equality $k(0) = \infty$ and assume that after this it will be locally summable). If μ is a nonnegative measure with compact support, then the *potential* of μ

with respect to the kernel k is the function u_μ,

$$u_\mu(t) = \int_G k(t-s)d\mu(s),$$

while the *energy* of μ (with respect to the same kernel) is the number $I(\mu)$,

$$I(\mu) = \iint_{G\times G} k(t-s)d\mu(s)d\mu(t).$$

Finally, the *capacity* of a compact set E, $E \subset G$, is the number $\operatorname{cap}_k E$,

$$\operatorname{cap}_k E = \sup\{\|\mu\|^2 : \operatorname{supp}\mu \subset E,\ I(\mu) \le 1\}. \tag{6}$$

The most important case is when $G = \mathbb{R}^n$ and k is very special, namely $k(t) = |t|^{-\alpha} \stackrel{\text{def}}{=} k_\alpha(t)$, $0 \le \alpha < n$, where $|\cdot|$ is the usual Euclidean norm (the kernels k_α are called the *Riesz kernels*; instead of $\operatorname{cap}_{k_\alpha}$ we write $\operatorname{cap}_\alpha$). In this case the notions just introduced are studied in an important and deep branch of analysis known as *potential theory*. Here we give just a rather sketchy survey of the connections with harmonic analysis, in particular with exceptional sets. First let us mention that the natural "exceptional" sets are now the sets of capacity zero. In general, estimating the capacity of a set with respect to appropriate kernels is a mean of appraising its "size".

Sets of capacity zero were already defined in Sect. 1 but not in the same way as here. These definitions are equivalent, provided k is a *kernel of positive type*, that is, it is the Fourier transform of a nonnegative function (the energy with respect to a measure μ is then given by

$$I(\mu) = \int |\hat{\mu}(u)|^2 \hat{k}(u)du,$$

and if $\hat{k} \ge 0$ we arrive at the definition in Sect. 1). In what follows we shall only consider kernels of potential type.[3] In particular, the Riesz kernels k_α are of this type.

3.2. Riesz Kernels. The *Newton kernel* k_2 in $\mathbb{R}^3$ and the corresponding potential have a well understood physical meaning (electrostatics, gravity). For detail we refer to Wermer's popular textbook (1974). It is with the study of the Newton kernel that potential theory began historically.

Let us state some fundamental facts of potential theory with the kernels k_α (the proofs can be found in Landkof (1966), Hayman and Kennedy (1976)). Let E be a compact subset of $\mathbb{R}^n$.

[3] If $k \notin L^1$, then the meaning of the condition $\hat{k} \ge 0$ has to be made more precise. If $G = \mathbb{R}^n$ one can invoke the theory of distributions. For general groups one may proceed, for example, as in Berg and Forst (1975).

Theorem 3.2.1. *There exists a unique measure μ for which the least upper bound in* (6) *is assumed with the kernel $k = k_\alpha$.*

This measure μ is called the *equilibrium distribution* on E (with respect to the kernel in question).

Theorem 3.2.2. *Let μ be the equilibrium distribution on E. Then the potential u_μ satisfies the inequality $u_\mu \leq 1$ everywhere on the support of μ, and $u_\mu \geq 1$ holds everywhere on E except on a set on which each measure of finite energy vanishes.*

It follows from the *maximum principle*, formulated below, that if $\alpha \geq n-2$ one has $u_\mu \leq 1$ everywhere (and not only on $\operatorname{supp} \mu$).

Theorem 3.2.3. (maximum principle). *If the inequality $u_\nu \leq M$ holds in the support of ν and if $n-2 \leq \alpha < n$, then $u_\nu \leq M$ everywhere.*

Theorem 3.2.3. (crude maximum principle). *If the same assumption on ν one has, in the case $0 < \alpha < n-2$, everywhere $u_\nu \leq 2^\alpha M$.*

Capacity is not an additive setfunction but it is subadditive: $\operatorname{cap}_\alpha(E_1 \cup E_2) \leq \operatorname{cap}_\alpha(E_1) + \operatorname{cap}_\alpha(E_2)$. If $n-2 \leq \alpha \leq n$ one has a stronger inequality ("convexity of capacity"):

$$\operatorname{cap}_\alpha(E_1 \cup E_2) + \operatorname{cap}_\alpha(E_1 \cap E_2) \leq \operatorname{cap}_\alpha(E_1) + \operatorname{cap}_\alpha(E_2). \tag{7}$$

A capacity can be extended to the system of all Borel (or even analytic) sets, cf. e. g. Landkof (1966). For $n-2 \leq \alpha \leq n$ such an extension is much easier to prove than for $\alpha < n-2$ (as inequality (7) is applicable in the former case, cf. *loc. cit.*). We say a few words about how the extension is done. The *inner capacity* (we fix, of course, the kernel beforehand) of an arbitrary set E is the number $\operatorname{cap}_* E = \sup\{\operatorname{cap} S : S \subset E, S \text{ compact}\}$. The *outer capacity* of E (notation: $\operatorname{cap}^* E$) is defined as $\operatorname{cap}^* E = \inf\{\operatorname{cap}_* U : U \supset E, U \text{ open}\}$. Finally, E is called *measurable* with respect to capacity, if $\operatorname{cap}^* E = \operatorname{cap}_* E$ (then $\operatorname{cap}^* E$ is called the capacity of E and is denoted simply $\operatorname{cap} E$). One can prove that all analytic sets are measurable with respect to capacity.

On the tori $\mathbb{T}^n$ one can construct a theory of capacities K_α. For instance, on the circle one can take as the analogue of the kernels k_α the functions $K_\alpha(e^{i\theta}) = |\sin \frac{\theta}{2}|^{-\alpha}$. (These functions blow up like $|\theta|^{-\alpha}$ at the origin, are 2π-periodic and have non-negative Fourier coefficients.) An importante rôle is played by the *logarithmic kernel* $K_0(e^{i\theta}) = -\log|\frac{\theta}{2}|$ (the corresponding capacity is written cap_0). In $\mathbb{R}^n$ one can likewise consider the logarithmic kernel $-\log|x|$ and the capacity generated by it – that the function $-\log|x|$ is nonnegative only for $|x| \geq 1$ does not cause much trouble.

3.3. Generalizations. We have seen that in the hypothesis $n-2 \leq \alpha < n$ the capacity $\operatorname{cap}_\alpha$ behaves better than when $\alpha < n-2$ (the maximum principle holds, and we have the convexity inequality (7)). This is connected with the fact that kernel k_α is subharmonic when $n-2 \le \alpha < n$. One can however find conditions which allow one to construct a capacity theory (including inequality (7)) without involving subharmonicity. In such a form it makes sense also for arbitrary groups.

Let G be a locally compact Abelian group and let ω a nonnegative locally summable function on the dual group Γ. By the *energy integral* of an (nonnegative) measure μ, $\mu \in G$, we intend the number $I(\mu)$,

$$I(\mu) = \int_\Gamma |\hat{\mu}(\gamma)|^2 \omega(\gamma) d\gamma$$

(here $d\gamma$ denotes the Haar measure on Γ). It turns out that one can build a rich theory of capacity corresponding to this energy integral provided the function ω^{-1} is continuous and *negative definite.* (If $\psi \ge 0$ then the condition of negative definiteness for ψ looks as follows: if $c_1, \dots, c_n \in \mathbb{C}$ and $c_1 + \dots + c_n = 0$ then

$$\sum_{1 \le i,j \le n} \psi(\gamma_j - \gamma_j) c_i \bar{c}_j \le 0$$

for arbitrary $\gamma_1, \dots, \gamma_n \in \Gamma$, cf. Berg and Forst (1975).)

Of course, it would be desirable to have the definition of energy, capacity and potential expressed in terms of a kernel K on G itself. It is clear that this kernel must be the inverse Fourier transform of the function ω. It is possible to attach to these words a strict sense – the kernel K will then in general be an infinite measure on G. The capacity will satisfy the convexity inequality (7) and, furthermore, a suitable analogue of the maximum principle (that is, Theorem 3.2.3) should hold. For details see, for example, Berg and Forst (1975).

However, from the analytic point of view it would be convenient if one had to deal with the case when the kernel K is a function and it would also be desirable if it satisfied some supplementary regularity conditions. In one word, it is necessary to have a reasonable balance between generality and easiness to handle the kernel at hand.

An example of a rather succesful balance is Preston's paper (1971).[4] There it is question of the case $G = \mathbb{T}$ (so that $\Gamma = \mathbb{Z}$ is the group of integers). We shall likewise confine ourselves to this framework, but it is, of course, possible to generalize everything (for example, to the case of $\mathbb{R}^n$).

Let $n \mapsto \lambda_n$ be a negative definite function on $\mathbb{Z}$ such that $0 < \lambda_0 \le \lambda_n$ and $\lambda_n = \lambda_{-n}$. One can prove that the assumption of negative definiteness is equivalent to the following:

(i) The function $n \mapsto (\lambda_n + \lambda)^{-1}$ is, for every $\lambda > 0$, the Fourier transform of a finite positive measure ν_λ.

[4] The reader should bear in mind that this paper has a technical defect, as noted by the reviewer in Math. Reviews 50.630, however not a very fundamental one.

We assume also that three more conditions are fulfilled:

(ii) The measure ν_0 is absolutely continuous with respect to Lebesgue measure. Its density K maps the circumference continuously into $[0, \infty]$ and $K(z) = \infty$ only for $z = 1$.

(iii) There exists a constant c_1 such that $K(e^{iy}) \leq c_1 K(e^{ix})$ for $0 < x < y \leq \pi$.

(iv) There exists a constant c_2 such that $K(e^{iy}) \leq c_2 K(e^{2ix})$ for $0 < x \leq \pi/2$.

As is shown in Preston (1971) for such a kernel (even without the condition (iv), which is needed for more subtle estimates; we will discuss this below) one can construct a theory of potential and capacity (denoted by cap_K) which is not less rich than for the Riesz kernels k_α in $\mathbb{R}^n$ with $\alpha \geq n-2$. On the other hand (cf. Preston (1971), Sect. 3) there are plenty of kernels satisfying the assumptions (i)–(iv). Namely, if the given sequence $\{\lambda_n\}$ can be presented in the form $\lambda_n = \lambda(|n|)$, where λ is a function in $C^3(\mathbb{R}_+)$ such that $\lim\limits_{x\to\infty} \lambda(x) = \infty$, $\lambda \geq 0$, $\lambda' \geq 0$, $\lambda'' \leq 0$, $\lambda''' \geq 0$ and $\lambda'(x) \leq c\,(1+x)|\lambda''(x)|$ then (i)–(iv) are fulfilled.

For example, one can take $\lambda(x) = (x+1)^\beta$, $0 < \beta < 1$, (then K behaves at the origin as $|t|^{-1+\beta}$) or $\lambda(x) = \log(x+2)$. Moreover, given any convergent series $\sum\limits_{n\geq 0} b_n$, $b_n \geq 0$, one can find a function λ subject to these conditions such that $\sum\limits_{n\geq 0} b_n\lambda(n) < \infty$.

With the kernel K (or the sequence $\{\lambda_n\}$) one can associate a Hilbert space H_K, consisting of those functions for which $\|f\|_K \stackrel{\text{def}}{=} (\sum\limits_{n\in\mathbb{Z}} |\hat{f}(n)|^2\lambda_n)^{1/2}$. (The energy of a measure μ with respect to K, of course, equals $\|u_\mu\|_K$, where u_μ is the potential of μ: $u_\mu(z) = \int K(z\bar{\zeta})d\mu(\zeta)$. The last statement in the foregoing paragraph assures that for each function in $L^2(\mathbb{T})$ there exists a kernel K satisfying (i)–(iv) such that $f \in H_K$.

3.4. Estimates for Maximal Functions. Convergence Quasi-Everywhere. The term "quasi-everywhere" means "everywhere except for a set of capacity zero". In this subsection we shall describe several classical situations in one dimensional Fourier analysis where questions of quasi-everywhere convergence are tightly intertwined with convergence almost everywhere. As this often is the case in convergence problems, it is convenient to formulate the main step in convergence proofs in terms of the corresponding maximal functions. Let us begin with two estimates proved in Preston (1971) but inspired by Carleson (1967) (where less general kernels are studied).

The *maximal function of Hardy-Littlewood* for a function f in $L^1(\mathbb{T})$ is defined by the formula

$$Mf(e^{it}) = \sup_{h>0} (2h)^{-1} \int_{|t-s|\leq h} |f(e^{is})| \frac{ds}{2\pi}.$$

The *maximal Hilbert transform* of f in $L^1(\mathbb{T})$ is defined as

$$H^* f(e^{it}) = \sup_{\varepsilon>0} \left| \int_{|t-s|\le\varepsilon} f(e^{is}) \cot\left(\frac{t-s}{2}\right) \frac{ds}{2\pi} \right| .$$

Theorem 3.4.1. *Assume that the kernel K satisfies the conditions* (i)–(iv). *Then there exists a constant $A = A(K)$ such that for every function f in H_K and every $\lambda > 0$ the estimate $\mathrm{cap}_K\{\zeta : Mf(\zeta) > \lambda\} \le A\lambda^{-2}\|f\|_K^2$ holds.*

Theorem 3.4.2. *Under the same assumptions on K one has the estimate* $\mathrm{cap}_K\{\zeta : H^*f(\zeta) > \lambda\} \le 16A\lambda^{-2}\|f\|_K^2$ *for every f in H_K and every $\lambda > 0$.*

It is well-known that if $f \in L^2$ then the Abel-Poisson means of the Fourier series of f converge a. e. to f (cf. for example the discussion of this question in the survey Kislyakov (1987) in this series). From Theorem 3.4.1 using standard reasonings (cf. Sect. 2.9 of Chapter 1 of the same survey) it is easy to see that if $f \in H_K$ then its Abel-Poisson means converge except on a set of K-capacity zero. Analogously, it follows from Theorem 3.4.2 that the limit of the "truncated" Hilbert transforms

$$\int_{|t-s|\le\varepsilon} f(e^{is}) \cot\left(\frac{t-s}{2}\right) \frac{ds}{2\pi}$$

exists except on a set of K-capacity zero provided $f \in H_K$, which may be viewed as an important complement to the theorem about a. e. existence of this limit for f in $L^2(\mathbb{T})$.

At this juncture it is proper to return to the end of the last subsection, where it was stated that every function in $L^2(\mathbb{T})$ lies in some space H_K. Thus, the approach via capacities indeed allows one to completely include the results on a. e. convergence of Abel-Poisson means and truncated Hilbert transforms for functions in $L^2(\mathbb{T})$. The set of divergence for each such function has measure zero, but in reality it is somewhat smaller – speaking very crudely, roughly as much as with which "excess" the series $\sum |\hat{f}(n)|^2$ converges.

The problem of the convergence of the partial sums $S_n f$, $S_n f = \sum_{|k|\le n} \hat{f}(k) z^k$, of the Fourier series of a function f, $f \in L^2(\mathbb{T})$, is much more difficult. In 1965 Carleson proved that $S_n f \to f$ a. e. (see the references in Kislyakov (1987)). However, the conjecture that this is so was made by N. N. Luzin several decades before. In the course of the years, when it was open, it did not lose the attention of mathematicians. Many results were established stating that at least something very close to *Luzin's conjecture* was true. One of the directions in which this was pursued was the problem of Weyl multipliers.

Let $\lambda_n > 0$ and assume that $\sum |\hat{f}(n)|^2 \lambda_n < \infty$. It is clear that if the numbers λ_n grow sufficiently fast (for example, as $|n|^a$, $a > 1$) then $S_n f \to f$ not only a. e. but even uniformly. Therefore it might be natural to ask whether one could not arrive at a proof (or a disproof?) of Luzin's conjeture by discovering the limit on the growth of the numbers λ_n where the a. e. convergence becomes divergence. If the condition $\sum |\hat{f}(n)|^2\lambda_n < \infty$ guarantees a. e. convergence then the numbers λ_n are said to be *Weyl multipliers.*

Before Carleson's theorem, which in many respects makes the very notion of Weyl multiplier obsolete, the best result here was the (now very old) theorem of Kolmogorov-Seliverstov, which states that the numbers $\lambda_n = \log(|n|+1)$ are Weyl multipliers. Its proof (it is an easy proof compared to the proof of Carleson's theorem) can be found in Zygmund (1959), Vol. 2, p. 161–162.

Later (but still long before the appearance of Carleson's theorem) it became clear that the condition $\sum |\hat{f}(n)|^2\lambda_n < \infty$ in fact must guarantee the convergence of the Fourier series quasi-everywhere with respect to a suitable capacity. Incidently, the case $\lambda_n = |n|^\beta$, $0 < \beta \le 1$, was mentioned in Kislyakov (1987): the corresponding capacity is $\mathrm{cap}_{1-\beta}$. The proof can be found, for example, in Kahane and Salem (1963). The reader may verify for himself that it is based essentially on the same computation as the Kolmogorov-Seliverstov theorem. The same idea can be used also in the derivation of Theorem 3.4.3 below. Thus, the "capacity point of view" is present as a shadow in the Kolmogorov-Seliverstov theorem.

Let us mention that in the case $\lambda_n = |n|^\beta$ the result just formulated is sharp: if $\mathrm{cap}_{1-\beta}E = 0$, where E is a closed set, then there exists a function f satisfying the condition $\sum |n|^\beta |\hat{f}(n)|^2 < \infty$ such that its Fourier series diverges on E. See, for example, Kahane and Salem (1963).

The case $\beta = 1$ plays a special rôle (then cap_0 is the logarithmic capacity), because a higher rate of growth in the power scale of the multipliers λ_n already guarantees absolute convergence of the Fourier series everywhere.

Let us now consider more general capacities. For a given kernel K set $\bar{K}(e^{it}) = |t|^{-1}\int_0^{|t|} K(e^{i\theta})d\theta$. Assume that also with respect to the kernel $\bar{K}$ one can construct a capacity $\mathrm{cap}_{\bar{K}}$ (in particular, $\bar{K} \in L^1(\mathbb{T})$) and that the function $t \mapsto K(e^{i\theta})$ is monotone on $[0, \pi]$.

Theorem 3.4.3. *In the above assumptions*

$$\mathrm{cap}_{\bar{K}}\{\sup_n |S_n f| < \lambda\} \le \mathrm{const}\,\lambda^{-2}\|f\|_K, \quad \lambda > 0$$

for all f *in* H_K.

For the proof see Carleson (1967) (more exactly, the derivation of formula (5.9) there). From this follows a general theorem about the convergence quasi-everywhere, which is also sharp in some supplementary assumptions on K, cf. Carleson (1967), Chapter 5.

We remark that the regularity assumptions imposed on K in Theorem 3.4.3 are considerably more stringent than those needed in the question of convergence of the Abel-Poisson means. They do not allow one to pass in the problem of Weyl multipliers beyond the limit $\lambda_n = \log(|n|+1)$ (if $K =$

$\sum_{n\in\mathbf{Z}} \lambda_n^{-1} z^n$ then for $\lambda_n = \log(|n|+1)$ the function $\bar{K}$ fails to be summable, although it is summable for $\lambda_n = [\log(|n|+1)]^{1+\delta}$, $\delta > 0$). Thus the capacity approach does not cover Carleson's theorem on a. e. convergence of the partial sums $S_n f$ (at least, not to the extent it is known today).

Concerning L^p analogues of the theorems of quasi-everywhere convergence, see Maz'ya and Khavin (1973).

3.5. Dimension. Hausdorff Measures. Let us return to the kernels k_α in $\mathbb{R}^n$. It is not hard to realize that if α grows then the set of measures with finite energy with respect to k_α shrinks. From this it follows that for a fixed E the function $\alpha \mapsto \operatorname{cap}_\alpha E$ is decreasing. The *capacitary dimension* of a (closed) set E is defined to be the number $\inf\{\alpha : \operatorname{cap}_\alpha E = 0\}$.

It turns out that it is possible to describe capacitary dimension in other terms connected with *Hausdorff measures*. Hausdorff measures can be defined for any metric space, but we shall confine ourselves to the space $\mathbb{R}^n$ (or $\mathbb{T}^n$).

Let h be a nondecreasing function mapping $[0, +\infty)$ into itself (in particular, h is continuous and $h(0) = 0$). For $E \subset \mathbb{R}^n$ and $\varepsilon > 0$ consider all possible coverings of E with countable families $\{B_n\}$ of balls with diameters not exceeding ε, and let $H(\varepsilon)$ be the infimum of all sums $\sum_n h(\operatorname{diam} B_n)$ taken over all these coverings. The number $\lim_{\varepsilon\to 0} H(\varepsilon)$ is called the h-Hausdorff measure of E and will be denoted $\operatorname{mes}_h E$. As is readily seen, the set function mes_h is an outer measure.

If $h_1(t) = o(h_2(t))$ for $t \to 0$ and $\operatorname{mes}_{h_2} E < \infty$ then, as is easy to see, $\operatorname{mes}_{h_1} E = 0$. A major rôle is played by the functions $t \mapsto t^\alpha$. The corresponding Hausdorff measure will be written $\operatorname{mes}_\alpha$. By what we just said it follows that for each set E there exists precisely one number β such that $\operatorname{mes}_\beta E = 0$ for $\beta > \alpha$ and $\operatorname{mes}_\beta E = \infty$ for $\beta < \alpha$. This number α is called the *Hausdorff dimension* of E and is denoted by $\dim E$.

If $\alpha = \dim E$ then there are three possibilities: $\operatorname{mes}_\alpha E = 0$, $0 < \operatorname{mes}_\alpha E < \infty$, $\operatorname{mes}_\alpha E = \infty$. In $\mathbb{R}^n$ one always has $\dim E \le n$.

Theorem 3.5.1 (Frostman). *The capacitary dimension and the Hausdorff dimension of a compact set coincide.*

The proof, and a detailed discussion of the material in the foregoing formulations can be found in Landkof (1966), Hayman and Kennedy (1980), Kahane and Salem (1963) (in the last book for the case of the unit circumference; of course, all that has been said above extends also to the n-dimensional tori $\mathbb{T}^n$, in particular, the circumference).

Let us state another theorem by Frostman (for its proof see again the above references). For us it is of interest, because it describes the class of sets of zero h-Hausdorff measure in the spirit of Sect. 1.5, namely, as the class of sets E which do not carry nonnegative measures μ with certain properties.

Theorem 3.5.2. *Let E be a compact set in $\mathbb{R}^n$ and assume that the function h satisfies the condition $h(2t) = O(h(t))$, $t \to 0$. Then $\operatorname{mes}_h E > 0$ if and only* *if there exist a nonnegative measure μ carried by E subject to the condition that $\mu(B \cap E) \leq h(\operatorname{diam} B)$ for each ball B.*

We remark also that in Kahane and Salem (1963) it is described how to compute the dimensions of some regularly structured perfect sets on the axis. In particular, the dimension of the Cantor ternary set is $(\log 2)/(\log 3)$.

3.6. Examples. Besides the capacitary estimates, estimates of the Hausdorff measure constitute one more way of deciding the "narrowness" of a set (if it is a question of a rather crude quantitative characteristic, namely the dimension, then both methods, as we have seen, give the same result). We give two examples (as well as the examples in Sect. 3.4, they are connected with the contents of the survey Kislyakov (1987)). They may also be viewed as illustrations of the theme "interrelation of classes of exceptional sets of different origin".

Theorem 3.6.1. *Let $0 < \alpha < 1$. There exist a probability measure μ on the circumference whose support is of vanishing α-Hausdorff measure, while its Fourier coefficients satisfy the inequality*

$$|\hat{\mu}(n)| \leq \text{const}\,(|n|^{\alpha} \log\log(|n|+1)^{-1/2}).$$

The point of the theorem is that it makes precise the connection of the rate of decrease of the moduli of the coefficients $|\hat{\mu}(n)|$ with the Hausdorff dimension of the set $\operatorname{supp}\mu$: a decrease with the rate of $|n|^{-\beta/2}$ with $\beta > \alpha$ is not possible, as the capacitary dimension and Hausdorff dimension coincide.

For the proof of Theorem 3.6.1 see Körner (1986). As is seen from the title of this paper, this result is connected with a theorem due to O. S. Ivashev-Musatov on the existence of measures, singular with respect to Lebesgue measure, the Fourier coefficients of which drop off with a prescribed rate. It is discussed in Kislyakov (1987). Besides, in Körner's paper (1986) just quoted one finds the following formulation of the Ivashev-Musatov theorem:

Let $\varphi : [0, \infty) \to [0, \infty)$ be a continuous positive function with $\int_1^{\infty} \varphi(x)^2 dx = \infty$ such that there exists a constant K, $K > 1$, such that $K\varphi(x) \leq \varphi(y) \leq K^{-1}\varphi(x)$ for $2x \geq y \geq x \geq 1$. Then there exists a measure μ on $\mathbb{T}$, $\mu \geq 0$, $\mu(\mathbb{T}) = 1$, supported on a set of zero Lebesgue measure, such that $|\hat{\mu}(n)| \leq \varphi(|n|)$ for $n \neq 0$.

In comparison with previous formulations the conditions on the function φ are here, apparently, reduced to a minimum.[5]

Theorem 3.6.1 is in fact given in a more general form in Körner (1986): instead of α-Hausdorff measure one considers h-measure with a function h which is not necessarily a power and the function φ enters in the formulation (precisely as above), but rather heavy regularity conditions are imposed on

[5] Cf. also Körner (1987)

h and φ. The estimate for the Fourier coefficients of μ has the form

$$|\hat{\mu}(n)| \leq \varphi(h(|n|^{-1})^{-1})[\log(h(|n|^{-1})^{-1})]^{1/2} \quad \text{for} \quad n \neq 0. \tag{8}$$

The regularity conditions mentioned are not fulfilled if $h(t)$ tends to zero rather slowly for $t \to 0$. In Körner (1986) there is however a result (Theorem 1.4) where no conditions whatsoever are imposed on h, but the estimate obtained for the Fourier coefficients is then not as strong as (8). If we completely renounce from the quantitative side of the question, we can formulate the result as follows: for each function h there exists a set of zero h-Hausdorff measure which is not a set of uniqueness in the narrow sense.

Sets of uniqueness in the wide and in the narrow sense were defined in Sect. 1.5 (among other examples of exceptional sets). We have already told that they are discussed in the survey Kislyakov (1987). There, among other things, we have quoted a theorem on the existence of a Helson set which is not a set of uniqueness (but each Helson set is a set of uniqueness in the narrow sense). Let us state a strengthening of this; cf. Körner (1973), p. 101.

Theorem 3.6.2. *For every function h and every $\alpha \geq 0$ there exists a Helson set on the circumference such that its Helson constant equals one but whose h-Hausdorff measure is α.*

References where to find other examples of applications of capacities and Hausdorff measures to the theory of exceptional sets (which theory, we recall, does not exist – the reader has probably by now found out this by himself) are contained in the Annotated Literature.

Annotated Literature

The books Carleson (1967) and Lindahl and Paulsen (1971) and Chapter 7 in Kahane (1968) are devoted to exceptional sets "in general", to which we may also add Körner's paper (1973), especially its Appendix. Carleson (1967) is mainly about convergence problems, sets of removable singularities and the like. The three other sources study the interdependence of some "lordly family" of classes of exceptional sets in harmonic analysis – cf. for example the tables in Kahane (1968), Chapter 7, Sect. 11 and Körner (1973), pp. 213–214. Let us remark that most of these classes follow the scheme set forth in Sect. 1.

References to Sidon sets are given in Sect. 2. To these we may further add Chapters V–VI in the book Lindahl and Paulsen (1971).

In Sect. 3 we have in passing mentioned the possibility to extend the capacity to a much larger system of sets. This result belongs to the so-called "axiomatic theory of capacity" (which is to some extent parallel to abstract measure theory and generalizes the latter). One can find information about this, for instance, in the books Dellacherie (1972a), (1972b); the latter gives also the connections to the theory of stochastic processes (the material of the book Berg and Forst (1975) has also, in essence, arisen from these connections).

Finally, let us list some books and articles, where one can find a variety of examples of the utilization of capacities and metric characteristics for the estimate of the "size" of exceptional sets arising in harmonic analysis and in various analytic disciplines: Maz'ya and Khavin (1973), Carleson (1967), Kahane (1968), (1970), (1985), Kahane and Salem (1963), Körner (1973), (1986), Preston (1971).

Bibliography*

Berg, C. and Forst, G. (1975): Potential theory on locally compact Abelian groups. (Ergeb. Math. Grenzgeb. 87). Berlin Heidelberg: Springer-Verlag (197 pp.), Zbl. 308.31001.

Bourgain, J. (1983): Sur les ensembles d'interpolation pour les mesures discrètes. C. R. Acad. Sci., Paris, Ser. I *296*, No. 3, 149–151, Zbl. 537.43016.

Bourgain, J. (1985): Sidon sets and Riesz products. Ann. Inst. Fourier *35*, No. 1, 137–148, Zbl. 578.43008.

Bourgain, J. (1989): Bounded ortogonal systems and the Λ_p-set problem. Acta Math. *162*, No. 314, 227–245, Zbl. 674.43004.

Bourgain, J. and V. Milman (1985): Dichotomie du cotype pour les espaces invariants. C. R. Acad. Sci., Paris, Ser. I *300*, No. 9, 263–266, Zbl. 584.43005.

Carleson, L. (1967): Selected problems on exceptional sets. Toronto London Melbourne: van Nostrand (151 pp.), Zbl. 189,109.

Dellacherie, C. (1972a): Ensembles analytiques. Capacités. Mesure de Hausdorff. Lect. Notes Math. 295. (123 pp.), Zbl. 259.31001.

Dellacherie, C. (1972b): Capacité et processus stochastiques. (Ergeb. Math. Grenzgeb. 67). Berlin Heidelberg: Springer-Verlag (103 pp.), Zbl. 246.60032.

Graham, C. C. and McGehee, O. S. (1979): Essays in commutative harmonic analysis. (Grundlehren Math. Wiss. 238). Berlin Heidelberg New York: Springer-Verlag (466 pp.), Zbl. 439.43001.

Hayman, W. K. and Kennedy, P. B. (1976): Subharmonic functions I. London New York San Francisco: Academic Press (284 pp.), Zbl. 419.31001.

Hewitt, E. and Ross, K. (1963/70): Abstract harmonic analysis I–II. (Grundlehren Math. Wiss. 115). Berlin Heidelberg New York: Springer-Verlag (519 pp.; 771 pp.), Zbl. 115,106, Zbl. 213,401.

Kahane, J. P. (1968): Some random series of functions. Lexington, Mass.: D. C. Heath & Co. (184 pp.), Zbl. 192,538.

Kahane, J. P. (1970): Séries de Fourier absolument convergentes. (Ergeb. Math. Grenzgeb. 50). Berlin Heidelberg: Springer-Verlag (169 pp.), Zbl. 195,76.

Kahane, J. P. (1985): Ensembles aléatoires et dimensions. In: Recent progess in Fourier analysis, North-Holland Math. Stud. *111*, 65–121, Zbl. 596.60075.

Kahane, J. P. and Salem, R. (1963): Ensembles parfaits et séries trigonométriques. Paris: Hermann (192 pp.), Zbl. 112,293.

Kislyakov, S. V. (1987): Classical themes of Fourier analysis. In: Itogi Nauki Tek. Ser. Sovrem. Probl. Mat. *15*, 135–195. VINITI: Moscow. English translation: Encyclopaedia Math. Sci. Vol. 15, pp. 113–165. New York: Springer-Verlag (1991), Zbl. 655.42006.

Körner, T. W. (1973): A pseudofunction on a Helson set I. Asterisque *5*, 3–224, Zbl. 281.43004.

Körner, T. W. (1986): On a theorem of Ivasev-Musatov III. Proc. London Math. Soc., III. Ser. *53*, No. 1, 143–192, Zbl. 618.42008.

Körner, T. W. (1987): Uniqueness for trigonometric series. Ann. Math., II. Ser. *126*, No. 1, 1–34, Zbl. 658.42013.

* For the convenience of the reader, references to reviews in Zentralblatt für Mathematik (Zbl.), compiled using the MATH database, and Jahrbuch über die Fortschritte der Mathematik (Jbuch) have, as far as possible, been included in this bibliography.

Landkof, N. S. (1966): Foundations of modern potential theory. Moscow: Nauka (515 pp.). English translation: (Grundlehren Math. Wiss. 180). New York Heidelberg: Springer-Verlag (1972), (424 pp.), Zbl. 148,103, Zbl. 253.31001.

Lindahl, L.-Å. and Paulsen, F. (eds.) (1971): Thin sets in harmonic analysis. (Lect. Notes Pure Appl. Math. 2). New York: Marcel Dekker (185 pp.), Zbl. 226.43006.

Lopez, J. M. and Ross, K. A. (1975): Sidon sets. (Lecture Notes Pure Appl. Math. 13). New York: Marcel Dekker (193 pp.), Zbl. 351.43008.

Marcus, M. B. and Pisier, G. (1981): Random Fourier series with applications to harmonic analysis. (Ann. Math. Studies 101). Princeton: Univ. Press (150 pp.), Zbl. 474.43004.

Marcus, M. B. and Pisier, G. (1984): Characaterizations of almost surely continuous p-stable random Fourier series. Acta Math. *152*, No. 3–4, 245–301, Zbl. 547.60047.

Mazya, V. G. and Khavin, V. P. (1973): Applications of (p, l)-capacity to some problems of the theory of exceptional sets. Mat. Sb., Nov. Ser. *90*, No. 4, 558–591. English translation: Math. USSR, Sb. *19*, 547–580 (1974), Zbl. 259.31007.

Pisier, G. (1977–78a): Ensembles de Sidon et espace de cotype 2. In: Sémin. Géom. des Espaces de Banach, Paris: Éc. Polytech. Exp. No. 14, 12 pp., Zbl. 416.46004.

Pisier, G. (1977–78b): Sur l'espace de Banach des séries de Fourier aléatoires presque surement continues. In: Sémin. Géom. des Espaces de Banach, Paris: Éc. Polytechn. Exp. No. 12/13, 33 pp., Zbl. 388.43009.

Pisier, G. (1983): Condition d'entropie et charactérisations des ensembles de Sidon. In: Topics in modern harmonic analysis, Proc. Semin., Torino Milano 1982, Vol. II, 911–944, Zbl. 539.43004.

Pisier, G. (1981): De nouvelles charactérisations des ensembles de Sidon. Adv. Math. Suppl. Stud. *73*, 685–726, Zbl. 468.43008.

Preston, C. J. (1971): A theory of capacities and its applications to some convergence results. Adv. Math. *6*, No. 1, 78–106, Zbl. 221.31006.

Rider, D. (1975): Randomly continuous functions and Sidon sets. Duke Math. J. *42*, No. 4, 759–764, Zbl. 345.43008.

Rudin, W. (1962): Fourier analysis on groups. New York: Interscience (285 pp.), Zbl. 107,96.

Varopoulos, N. Th. (1976): Une remarque sur les ensembles de Helson, Duke Math. J. *43*, No. 2, 387–390, Zbl. 345.43004.

Wermer, J. (1974): Potential theory. Lect. Notes Math. 408, (145 pp.), Zbl. 297.31001, 2nd ed. 1981 (166 pp.), Zbl. 446.31001.

Zygmund, A. (1959): Trigonometric series I–II. Cambridge: Univ. Press (383 pp.; 354 pp.), Zbl. 85,56.

Author Index

Subject Index

Encyclopaedia of Mathematical Sciences

Editor-in-chief: R. V. Gamkrelidze

Dynamical Systems

Volume 1: **D. V. Anosov, V. I. Arnol'd** (Eds.)

Dynamical Systems I

Ordinary Differential Equations and Smooth Dynamical Systems

1988. IX, 233 pp. 25 figs. ISBN 3-540-17000-6

Volume 2: **Ya. G. Sinai** (Ed.)

Dynamical Systems II

Ergodic Theory with Applications to Dynamical Systems and Statistical Mechanics

1989. IX, 281 pp. 25 figs. ISBN 3-540-17001-4

Volume 3: **V. I. Arnol'd** (Ed.)

Dynamical Systems III

1988. XIV, 291 pp. 81 figs.
ISBN 3-540-17002-2

Volume 4: **V. I. Arnol'd, S. P. Novikov** (Eds.)

Dynamical Systems IV

Symplectic Geometry and its Applications

1989. VII, 283 pp. 62 figs.
ISBN 3-540-17003-0

Volume 5: **V. I. Arnol'd** (Ed.)

Dynamical Systems V

Bifurcation Theory and Catastrophe Theory

1991. Approx. 280 pp. 130 figs.
ISBN 3-540-18173-3

Volume 6: **V. I. Arnol'd** (Ed.)

Dynamical Systems VI

Singularity Theory I

1992. Approx. 250 pp. ISBN 3-540-50583-0

Volume 16: **V. I. Arnol'd, S. P. Novikov** (Eds.)

Dynamical Systems VII

Nonholonomic Dynamical Systems. Integrable Hamiltonian Systems

1992. Approx. 290 pp. ISBN 3-540-18176-8

Commutative Harmonic Analysis

Volume 15: **V. P. Khavin, N. K. Nikol'skij** (Eds.)

Commutative Harmonic Analysis I

General Survey. Classical Aspects

1991. IX, 268 pp. 1 fig. ISBN 3-540-18180-6

Volume 25: **N. K. Nikol'skij** (Ed.)

Commutative Harmonic Analysis II

Group-Theoretic Methods in Commutative Harmonic Analysis

1992. Approx. 300 pp. ISBN 3-540-51998-X

Volume 72: **V. P. Khavin, N. K. Nikol'skij** (Eds.)

Commutative Harmonic Analysis III

Volume 42: **V. P. Khavin, N. K. Nikol'skij** (Eds.)

Commutative Harmonic Analysis IV

Harmonic Analysis in $\mathbb{R}^n$

1991. Approx. 240 pp. ISBN 3-540-53379-6

Encyclopaedia of Mathematical Sciences
Editor-in-chief: R. V. Gamkrelidze

Analysis

Volume 13: **R. V. Gamkrelidze** (Ed.)
Analysis I
Integral Representations and Asymptotic Methods
1989. VII, 238 pp. 3 figs.
ISBN 3-540-17008-1

Volume 14: **R. V. Gamkrelidze** (Ed.)
Analysis II
Convex Analysis and Approximation Theory
1990. VII, 255 pp. 21 figs. ISBN 3-540-18179-2

Volume 26: **S. M. Nikol'skij** (Ed.)
Analysis III
Spaces of Differentiable Functions
1991. VII, 221 pp. 22 figs.
ISBN 3-540-51866-5

Volume 27: **V. G. Maz'ya, S. M. Nikol'skij** (Eds.)
Analysis IV
Linear and Boundary Integral Equations
1991. VII, 233 pp. 4 figs.
ISBN 3-540-51997-1

Volume 19: **N. K. Nikol'skij** (Ed.)
Functional Analysis I
Linear Functional Analysis
1992. Approx. 300 pp. ISBN 3-540-50584-9

Volume 20: **A. L. Onishchik** (Ed.)
Lie Groups and Lie Algebras I
Foundations of Lie Theory. Lie Transformation Groups
1992. Approx. 235 pp. ISBN 3-540-18697-2

Several Complex Variables

Volume 7: **A. G. Vitushkin** (Ed.)
Several Complex Variables I
Introduction to Complex Analysis
1990. VII, 248 pp. ISBN 3-540-17004-9

Volume 8: **A. G. Vitushkin, G. M. Khenkin** (Eds.)
Several Complex Variables II
Function Theory in Classical Domains. Complex Potential Theory
1992. Approx. 260 pp. ISBN 3-540-18175-X

Volume 9: **G. M. Khenkin** (Ed.)
Several Complex Variables III
Geometric Function Theory
1989. VII, 261 pp. ISBN 3-540-17005-7

Volume 10: **S. G. Gindikin, G. M. Khenkin** (Eds.)
Several Complex Variables IV
Algebraic Aspects of Complex Analysis
1990. VII, 251 pp. ISBN 3-540-18174-1

Volume 54: **G. M. Kenkin** (Ed.)
Several Complex Variables V
1992. Approx. 280 pp. ISBN 3-540-54451-8

Volume 69: **W. Barth, R. Narasimhan** (Eds.)
Several Complex Variables VI
Complex Manifolds
1990. IX, 310 pp. 4 figs. ISBN 3-540-52788-5

GPSR Compliance

The European Union's (EU) General Product Safety Regulation (GPSR) is a set of rules that requires consumer products to be safe and our obligations to ensure this.

If you have any concerns about our products, you can contact us on ProductSafety@springernature.com

In case Publisher is established outside the EU, the EU authorized representative is:

Springer Nature Customer Service Center GmbH
Europaplatz 3
69115 Heidelberg, Germany

Batch number: 10620141

Printed by Printforce, the Netherlands